VALUE STREAM MAPPING
FOR THE
PROCESS INDUSTRIES

VALUE STREAM MAPPING

FOR THE

PROCESS INDUSTRIES

Creating a Roadmap for Lean Transformation

Peter L. King
Jennifer S. King

CRC Press
Taylor & Francis Group
Boca Raton London New York

CRC Press is an imprint of the
Taylor & Francis Group, an **informa** business

A PRODUCTIVITY PRESS BOOK

CRC Press
Taylor & Francis Group
6000 Broken Sound Parkway NW, Suite 300
Boca Raton, FL 33487-2742

© 2015 by Taylor & Francis Group, LLC
CRC Press is an imprint of Taylor & Francis Group, an Informa business

No claim to original U.S. Government works

Printed on acid-free paper
Version Date: 20150224

International Standard Book Number-13: 978-1-4822-4768-8 (Paperback)

Library of Congress Cataloging-in-Publication Data

King, Peter L.
    Value Stream Mapping for the process industries : creating a roadmap for lean transformation / Peter L. King, Jennifer S. King.
        pages cm
    Summary: "This book describes in detail how to create a complete Value Stream Map for a process industry manufacturing operation. Describing the unique features of process operations and why they require additions and adjustments to traditional VSMs, the book describes an actual process operation and uses it to describe in detail how to create the VSM. It defines all of the terms required for process VSM data boxes and uses the example process to illustrate all of the calculations required"-- Provided by publisher.
    Includes bibliographical references and index.
    ISBN 978-1-4822-4768-8 (paperback)
    1. Manufacturing processes. 2. Production management. 3. Value analysis (Cost control) 4. Process control. I. King, Jennifer S., 1980- II. Title.

TS183.K563 2015
658.5--dc23
                                                                                        2014042965

Visit the Taylor & Francis Web site at
http://www.taylorandfrancis.com

and the CRC Press Web site at
http://www.crcpress.com

Contents

Acknowledgments

As the lead author of this work, I am first and foremost grateful to my daughter Jennifer for agreeing to join me in this endeavor. Her help in developing the text, and in editing it and bringing a higher level of clarity, made this task much easier and a real joy. This is our second book together, and I'm finding her collaboration leads to a much better product.

Neither the concepts, thoughts, ideas, and the guidance presented in this handbook, nor the experience in their practical application, are solely my work. I am indebted to a number of colleagues who have helped me travel this journey:

- Bill Sheirich, Vinay Sohoni, and John Anderson, IBM consultants who first exposed me to IBM's Line Analysis Mapping, a 1980s forerunner to VSM.
- Wayne Smith, manager of the DuPont Continuous Flow Manufacturing (CFM) collaboration with IBM, who recognized the need to add information flow to the Line Analysis format.
- Bennett Foster, who helped me map several of DuPont's operations, including one that manufactured sheet goods for use in weatherproofing houses and another that laminated circuit board substrates. Bennett's contributions included the development of flow simulations based on the VSMs, which were invaluable in scoping improvements and designing pull systems to resolve wastes discovered through the maps.
- Ted Brown, who pioneered the use of VSM concepts and formats to analyze DuPont's global supply chains.
- My colleagues in DuPont's Lean Center of Competency: Paul Veenema, Cris Leyson, Laura Colosi, Nick Mans, Larry Jordan, Terry Farr, John New, Larry Mlinac, Lew Buckminster, and Paul Jungling, all of whom contributed to the body of knowledge and experience described here.
- Colleagues in DuPont business units who championed the use of Value Stream Mapping within their respective businesses: Pete Ellefson, Ed Reiff, Anne Kraft, Karen Wrigley, John Rees, Tom Carroll, Mike Archie, and Fran Montemurro.
- Michael Sinocchi, Executive Editor at Productivity Press, for his ongoing support, guidance, and advice for this and for my prior literary efforts. It was Mike's encouragement that led me to write my first book, which opened up a whole world of new possibilities and enriched my life.

- My daughter Courtney, herself a professional in the field of operations management and production planning and forecasting, for providing insights and experiences that were invaluable to this effort.
- Finally, to my wife Bonnie who not only provided strong emotional support and encouragement, but also takes care of managing the affairs of Lean Dynamics so that I have time to do the fun stuff like working face-to-face with clients and writing this book.

Introduction

Why This Book Is Important

A Value Stream Map (VSM) is a process flow chart that shows each step in the production of a good or material, as well as the resources used in each step, and the relationships between the resources. Value stream mapping is a very significant component of any Lean initiative, providing a framework that highlights waste and the negative effect it has on overall process performance and flow. The format proposed in *Learning to See,* having evolved from Toyota's material and information flow maps, has become the de-facto standard for VSM. However, while *Learning to See* is a landmark work, it is written primarily from a discrete parts production and assembly point of view, using the manufacture of steering components for tractors as an example to describe value stream mapping. The book doesn't nearly describe all of the issues encountered in mapping a process industry operation or ways to see some of the unique wastes inherent in a process operation.

While this conventional VSM format has the basic structure to effectively describe process operations, it must be adapted and expanded to serve its purpose in that environment. By process operations, we mean those characterized by chemical and mechanical transformations, including mixing, blending, chemical reactions, extrusion, sheet forming, slitting, baking, and annealing. Finished products can be in solid form packaged as rolls, spools, sheets, or tubes; or in powder, pellet, or liquid form in containers ranging from bottles and buckets to tank cars and railcars. Examples include automotive and house paints, processed foods and beverages, personal care products, paper goods, plastic packaging films, fibers, carpets, glass, and ceramics. The output may be sold as consumer products such as breakfast cereal, salad dressing, shampoo, toothpaste, or brake fluid, but more often become ingredients or components for other manufacturing processes.

The production of these materials differs from the manufacture and assembly of discrete parts in a number of significant ways:

■ The process steps involve chemical reactions like mixing, blending, and polymerization, and mechanical transformations like stretching, annealing, extrusion, and casting.

■ Process operations tend to be very equipment intense rather than labor intense, and throughput is most often limited by equipment rather than labor. Therefore, a process VSM must focus more on wasted capacity than on wasted labor (movement, waiting, etc.). It must include data that more clearly describe equipment performance and highlights capacity waste, such as OEE (overall equipment effectiveness) and its components: quality, reliability, and rate.

■ Material types or product types tend to increase as material goes through the sequential process steps, as more and more product differentiation occurs. (In assembly processes, the number of individual part types tends to decrease as parts are assembled.) Therefore, the material type fan out must be shown to get an understanding of flow dynamics.

■ In contrast with assembly operations, which deal with discrete physical things that can be counted, process operations deal with materials that generally change form as they move through the process. Therefore, careful consideration must be given to the most appropriate units to be used in the data box for each process step. For example, in a salad dressing bottling operation, material could be measured in pounds, gallons, bottles, cases, or pallets. As another example, the output from a plastic film extrusion line could be measured in pounds, lineal feet, square feet, or rolls. In addition, the relationships between these parameters often vary with product type (two 2000-foot rolls of plastic packaging film may have different weights because of different film thickness or film width). Therefore, the units used in the VSM data boxes must be carefully selected to most clearly illustrate flow and bottlenecks.

■ Product changeovers (aka product transitions, setups) often include material losses as well as time losses, so this aspect of waste must be shown.

Some of the guidance typically given in mapping handbooks doesn't work well for a process operation. Some of it is actually wrong for processes, and will cause errors in analysis and decision making. Lean authorities often speak of separating the plant flow into value streams, and treating each value stream as a separate entity, but in many process operations, all value streams flow through shared equipment and so they are all interdependent from a flow point of view. Thus a single VSM which includes all value streams is much more descriptive than a VSM that focuses on a single value stream. Narrowing to a single product family can understate Takt (the Lean term for customer demand) and hide flow issues and bottlenecks.

How the Book Is Organized

The book introduces a target manufacturing process, describes it in enough detail that all information needed for a complete VSM is provided, and then walks through the entire procedure for generating the map. The target process is complex enough to illustrate the issues often encountered in mapping a process industry operation, but still straightforward enough to explain all of the mapping considerations and decisions within a book of this length. The book also describes how to analyze the map for wastes and flow issues so that they can be reduced or eliminated to lead toward an improved future state.

- Chapters 1 through 4 lay out how to create a VSM for a process industry operation, explain how it differs from and expands on conventional value stream mapping, and describes best practices for creating a VSM.
- Chapter 5 describes the target process, gives all the data required to map it, and describes the scheduling and communication processes that form the basis for the information flow portion of the map.
- Chapters 6 through 11 walk you through the detailed steps in creating a VSM for the target process, including examples of all the calculations needed for the flow parameters in the data boxes
- Chapters 12 through 15 tell you how to analyze the map to better understand process flow and wastes, how to scope and prioritize opportunities, and how to create a multi-generational set of future state VSMs for the target process.
- Chapter 16 describes how the VSM concept can be applied to entire supply chains, and why a Supply Chain Map (SCM) is as important to supply chains as a VSM is to manufacturing operations.
- Chapters 17 and 18 describe how to engage the entire workforce, indeed the entire organization, in map creation and its ongoing use as a roadmap for continuous improvement.
- Finally, Chapter 19 gives a few examples of the use of VSMs in real processes, and describes situations where VSMs were valuable in helping operations make the right decision or avoid the wrong one.
- There are several appendices to explain in detail how to calculate some of the data box parameters and to describe some Lean tools often used to reach future state.

The book is structured in a way that meets the interests of a wide variety of readers. If you are looking for an in-depth walk-through of the details of generating a VSM, analyzing it to recognize opportunities to improve performance, and then sort, prioritize, and illustrate them on a future state VSM, that is all covered here. If, on the other hand, you only want an understanding of why a process VSM must be different from a traditional VSM and then the broad concepts of how to create that type of VSM, you can get that by focusing on

Chapters 1 through 4, giving Chapters 6 through 12 a light reading, and skipping Chapters 13, 14, and 15.

Roadmap for a Complete Lean Transformation

The purpose of this book is to explain how to create a VSM for a process operation, and to walk you through the steps in analyzing the map, scoping improvement projects, prioritizing them, and then using future state VSMs to illustrate and motivate systemic improvement. In doing so, it actually provides a template or roadmap for a complete Lean transformation. It describes all the Lean initiatives to be undertaken to reduce four of the seven classic wastes, the four most often found in process operations. It describes projects to eliminate overproduction, reduce inventories, shorten cycle times, reduce yield losses, implement cellular manufacturing level production using a form of heijunka called product wheels, and implement a pull replenishment system. It ties all these improvements together in a coordinated, integrated architecture so that improvements are done in the right sequence and in a way that they build upon each other.

By following the guidance given within this handbook, you will be able to construct a complete VSM for your process operation. Using the insights provided by that map, you can see more clearly the wastes in your operation, understand their root causes, and begin to move toward a much improved future state.

I hope that as you read this that it becomes apparent that a well-constructed, high-quality VSM provides a template and a roadmap to guide your Lean transformation.

Chapter 1

The Value of Mapping

Reducing waste and improving material flow are two of the primary goals of Lean. In order to reduce or eliminate waste, you have to understand where it exists in your process, and a value stream map (VSM) is designed to enable you to do just that. Toyota developed its material and information flow maps, on which the VSM format is based, specifically to see waste and its causes.

Lean defines waste as anything that consumes resources (material, people, and equipment) but doesn't create value for the customer. Toyota defined seven types of waste:

1. Overproduction—making more than the customer needs, or making it sooner than needed
2. Inventory—material not currently being processed, including raw materials, work in process, and finished product inventory
3. Defects—parts or material that do not meet required specifications
4. Transportation—movement of material, either from one process step to the next step or into or out of inventory
5. Waiting—time that operators, mechanics, or anyone else spend waiting for material or for the equipment to be ready to use
6. Movement (of people)—walking around the equipment to get where they are needed, or to get changeover parts or tools
7. Processing—excessive processing, doing more to the material than the customer requires

The first four of these can be readily seen from a well-constructed VSM. Interestingly, these first four are generally the most prevalent and most expensive wastes in process operations. Therefore, a VSM can illustrate the wastes of highest concern to those responsible for process performance. The remaining three require more detailed analysis, using Lean methods such as movement charts called "spaghetti diagrams." An eighth waste is often added to Toyota's seven—the waste of human knowledge, creativity, and potential. This also is something

1

that can't readily be seen on a VSM; it requires a thorough analysis of work-place culture, attitudes, behaviors, and participation in continuous improvement processes.

A VSM will also illustrate end-to-end flow across your operation, show how value is being created for your customers, and clarify barriers to smooth flow. These are often difficult to see in a process operation because many of the value adding steps take place in tanks, vessels, and piping, and the size and complexity of those make process flow far less visible.

Once the wastes and flow barriers are understood, and plans to eliminate them have been formulated, most of the data needed to estimate benefits of these improvements can be found on the VSM. The VSM can then be used as a template to define the future state that would result from successful implementation of those plans. Thus, it creates and quantifies a vision that can be used to motivate action toward the future state.

A VSM is at a high enough level that you can see flow through the entire process, and thus create a strong cross-functional view of the whole process. Figure 1.1 is an example of a well-constructed VSM, one that we will be revisiting throughout this book.

BENEFITS OF A VALUE STREAM MAP
It provides an understanding of the operations that are creating value for the customer.
It provides a clear view of material flow from raw materials to finished products, and makes barriers to smooth flow very visible.
It provides an integrated picture of the process, which improves understanding of the interactions between various steps.
If the VSM is created by a team representing all process areas and all functions, it builds a strong cross-functional understanding of the overall process and its interconnectedness.
It highlights the main forms of waste found in a process operation.
It provides some clues to the root causes of the wastes.
It ties information flow to material flow, so that the effect of errors, delays, and re-work in information processing on smooth material flow becomes visible.
It provides a template for the design of an improved future state, and the data to quantify performance improvements.
It provides a sound basis for chartering Kaizen Events, with confidence that they are focused on problems affecting process performance.

A Focus on Flow Rather Than on Function

Plants have tended to be managed on a functional area-by-area basis, with area-based supervision and area-based goals and metrics, and little plant-wide focus. This is especially true of process plants; a chemical plant traditionally may be divided into a polymerization area, an extrusion area, a compounding area, a pelletizing area, and a finishing and packaging area, with no integrated view of

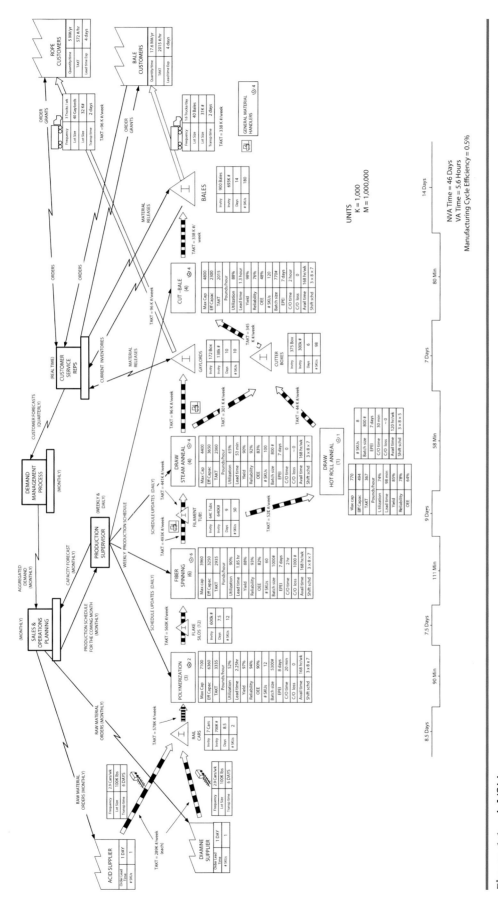

Figure 1.1 A VSM.

overall flow. A VSM helps to see process interactions and the interconnectedness of sequential process steps.

It also clarifies how the information processing activities influence material flow and operational performance, and how dysfunction in information handling can impede material flow and degrade process performance.

Summary

A properly drawn VSM is an extremely important component of any Lean activity. It provides a detailed understanding of the current state, in a way that clarifies flow and detractors to smooth flow. It accurately depicts the major effects of waste and wasteful processes, and provides insight into root causes of waste. It is the starting point for creation of a vision of what the future state should look like—the future state VSM. It provides a template and design information for application of Lean improvements like cellular manufacturing, production leveling, and pull replenishment systems. It provides a context for prioritizing all of the improvement initiatives arising from analysis of the VSM. In addition, as explained in the Introduction, it can provide the architecture for a complete Lean transformation of your operation. It should be the first technical work activity in any Lean transformation, begun as soon as the organizational work and team formation are completed.

Chapter 2

Value Stream Mapping Fundamentals

Introduction to Value Stream Mapping

In the last chapter, we mentioned the various types of waste, and that a VSM is an effective way to highlight waste and its causes. We also pointed out that a well-constructed VSM illustrates flow and the barriers to smooth flow, things that cause longer lead times and higher inventories.

For most process industry operations, the format described in works like *Learning to See* provides a good starting point. However, to describe and understand the additional complexities inherent in many process industry plants, additional features and data are often necessary. A slightly different approach to the creation of the map is often warranted. However, before introducing those features and approaches, let's look at traditional VSM composition.

A VSM consists of three main components as shown in Figure 2.1:

1. Material flow—Shows the flow of material as it progresses from raw materials, through each major process step (machine, tank, or arrangement of vessels), to finished goods moving toward the customer. This is a high-level view showing only major pieces of equipment or processing systems, with data boxes that illustrate the performance of each piece. All inventories along the flow are also shown, with data boxes that show the contents of each inventory storage.
2. Information flow—The flow of all major types of information that govern what is to be made and when it is to be made. This starts with orders from the customer, traces back through all significant planning and scheduling processes, and ends with schedules and control signals to the production floor. Information typically flows in the opposite direction of material flow.

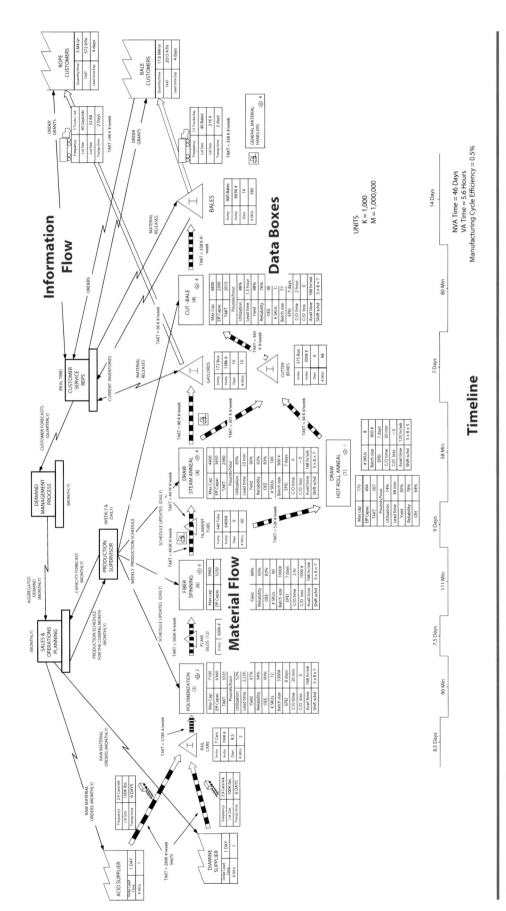

Figure 2.1 A complete VSM and its components.

3. Timeline—Shows the value-add time and contrasts it with non-value-add time. It is a line at the bottom of the VSM in the form of a square wave. This is a key indicator of waste in the process: it shows the effect of waste but not the cause; that should be diagnosed from the other two components of the VSM.

Material Flow

Major Process Steps

Each major step in the process will be described on the VSM by a process box. A process box may depict a single large machine or chemical process, like a carpet-tufting machine, a paint-mixing vessel, or a bottle-filling machine in a salad dressing plant. It could also depict a step in the process consisting of an integrated process system; a carpet dyeing system consisting of dye mix tanks, heaters, pumps, and the dye application machine, or a continuous chemical polymerization system with several tanks and much process piping, could each be shown as a single process box.

The number of operators normally working at this process step is shown within the process box. If people are shared between steps, fractional values may be shown. If, for example, a step has a dedicated operator and an operator shared with another process step, it could be shown as 1.5 operators for this step. In assembly processes, this can be one of the most important parameters on the VSM; in the process industries, it is usually far less important than equipment utilization, which is one of the parameters listed in the data box.

Data Boxes

Data boxes provide the numerical information required to understand how well material is flowing through the process, where bottlenecks or capacity constraints exist, where waste exists in the process, and to provide clues to the root causes. Data boxes quantify not only the waste, but also the effects of the waste, such as inventories caused by costly set ups or by overproduction.

The five types of data box normally shown on a VSM are described next. The typical make-up of each type is shown; for some specific situations, additional data may be relevant and should be included in the data box. There may also be cases where not all of the data listed is important; in those cases, the data box should be shortened for simplicity. The intent is to show enough of the operational details to give a clear understanding of flow and its influencing factors, but not to overwhelm the map with unneeded data. It should also be emphasized that extreme accuracy is not the goal; in most cases, reasonable approximations are sufficient. Where values can vary significantly, the range is often listed.

All of the data listed next that is relevant to the specific process being mapped should be shown on the VSM. The data should include all parameters that influence flow; it should not include data related to the chemical or mechanical processing, like temperatures, pressures, viscosities, or torque requirements. In some cases, the data won't be enclosed in a box, but simply be shown as text along a flow line. This is sometimes done with transportation data, for example.

Process Box and Process Data Box

The parameters normally included in a process step data box (Figure 2.2) are listed here. Chapter 7 gives details on how to calculate each parameter with examples.

■ Maximum Capacity—The throughput you could expect from this process step under "perfect" conditions, that is, the maximum throughput when there are no yield losses, no rate reductions, no failures or momentary upsets, and no time spent in changeovers. This is sometimes called name-plate capacity or maximum demonstrated capacity.

■ Effective Capacity—This is the throughput you can expect when things are running normally, with the average yield losses and the typical maintenance and changeover downtimes. It can be listed on an hourly or daily basis or whatever time frame makes sense for this process. It is calculated by multiplying Maximum Capacity by Overall Equipment Effectiveness (OEE), defined below. It is a realistic measure of what you can expect this step to produce over reasonably long periods, under current performance.

■ Takt—A measure of customer demand. It is the total aggregate demand placed on this step divided by the available time. Takt defines what a step in the process must produce, and Effective Capacity defines its ability to do that, so Takt should be listed in the same time units as Effective Capacity.

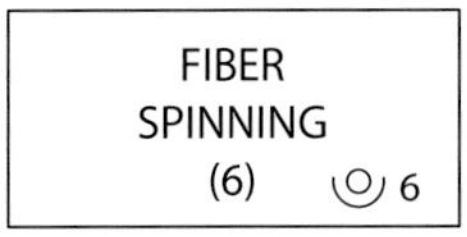

FIBER SPINNING (6) ↻ 6	
Max cap	3960
Eff Capac	3250
TAKT	2935
Ponds/hour	
Utilization	90%
Lead time	1.85 hr
Yield	88%
Reliability	93%
OEE	82%
# SKUs	60
Batch size	1000#
EPEI	7 days
C/O time	2 hr
C/O loss	1000 #
Avail time	168 hr/wk
Shift schd	3 × 8 × 7

Figure 2.2   Process box and process data box.

Takt is one of the most important parameters to appear on a VSM. The term comes from a German word meaning rhythm, or drum beat. The goal is to synchronize every part of the manufacturing operation to the rhythm of customer demand, so that customer demand can be fully met while avoiding the waste of overproduction. In traditional Lean, it is a time-based factor calculated by taking the time available, the total time the plant plans to be operating over some period, and dividing it by the average number of units of product that customers purchase over that time period. The fact that Takt is a time-based parameter often causes confusion in process industry applications. Operators, supervisors, and process engineers are much more accustomed to dealing with rate parameters. Stating throughput as 5000 gallons per hour means much more to them than 0.72 seconds per gallon. The two are mathematically equivalent, the rate values being simply the reciprocal of the time parameters. More will be said about this in the next chapter. For all of the examples in this book, we will calculate Takt as a rate parameter rather than as a time parameter.

It is important to recognize that Takt may sometimes be different at each step in an operation. If there are scrap or yield losses inherent in downstream processes, the Takt for the upstream processes must be increased to accommodate those losses. If, for example, step 4 in a process has a 10% yield loss, then steps 1, 2, and 3 have to produce 10% more material to make up for the loss. When we say that Takt is a measure of customer demand, that is not necessarily the final customer; it may be an "internal customer," that is, the next step in the process.

Takt may also be different from step to step if different areas run different shift schedules. It is sometimes the case in process plants that different areas have different available times. The bottleneck operations generally run 24/7, but non-bottlenecks may run a reduced schedule. Bagging, wrapping, and packing areas, for example, may run only an 8-hour, 5-day schedule. Since the time available can vary from step to step, so will the Takt. In a film coating process, there may be two coaters in parallel, with significantly different processing capability. If demand for the products requiring Coater 1 is significantly less than the demand on Coater 2, Coater 1 might be run only two shifts per day, with Coater 2 running three shifts. Thus, the Takt for the two coaters would differ because of different demand levels and different available times.

Since available time does not include time the equipment is down for lunch periods and breaks, Takt will differ between areas that do and do not continue to operate during lunch and breaks.

■ Utilization—This is a measure of how fully utilized a process step is, and provides a key to how close to a bottleneck it may be. It is calculated as Takt rate divided by Effective Capacity, that is, how much material you need divided by how much you can produce. Since many process industry production lines are asset limited, this parameter highlights which specific steps are critical to throughput. Any step with a utilization of 100% is a bottleneck, and a step at 90% could easily become a bottleneck under some operating conditions. A utilization of 85% or less is considered "comfortable" because utilization is based on effective capacity, which accounts for normal losses and downtime.

■ Lead Time—The time it takes for one part, one batch, one roll, one tote, or one lot to complete that process step, from the time the material enters the process step until it leaves. The lead time may include both value-add time, where a chemical or mechanical transformation for which the customer has value is being performed, and non-value-add time, that is, wasted time.

■ Yield—The percentage of the material entering the step which leaves with all properties in acceptable ranges for all downstream processing.

■ Reliability—The percentage of time that the equipment is *not* down because of mechanical, electrical, or control system equipment failure, that is, 100% minus the percentage of downtime.

■ OEE—A holistic measure of equipment performance; a metric that encompasses all forms of time lost: reliability downtime, Preventive Maintenance (PM) tasks, yield losses, set-up or campaign changeover time, rate reductions, etc. Some process companies call this Uptime, and the calculation can be slightly different, but the same parameters are included and the numerical result is the same. OEE is explained more fully in Chapter 7.

■ # SKUs (Stock Keeping Units)—The number of specific product types or materials leaving this process step. The number should account for all differentiating features. Where the data boxes show many more SKUs leaving a process step than entering, that signifies a highly differentiating step. If a process step has equipment in parallel, and specific products require specific machines or vessels, the SKU count for each should reflect the portion of the total SKU number processed on that machine or vessel.

■ Batch Size—The amount of material produced as a single lot or batch. Examples of a single batch are a single roll of film, paper, or carpet; a quantity of dough made as a unit, although it may exit this step as a number of loaves; a mixing vessel's quantity of paint; a reactor's quantity of bulk chemicals. This should not be confused with campaign size, where several batches of one SKU may be run before changing to produce another SKU.

■ EPEI—A Lean term for "Every Part Every Interval." It is the time span over which all regularly produced product types are made. There are often cases where low volume products are not made every interval; in those cases, EPEI would be the time span over which all of the high volume products are made. If product wheels are the scheduling methodology used, EPEI is equal to wheel time.

■ Changeover (C/O) Time—The time to change from one product type to another, including the time to get to full rate on the new product and get all properties within quality specifications. Sometimes referred to as "good product A to good product B time." Also called set-up time or transition time. If different products made on a piece of equipment have different changeover times, a weighted average should be shown.

- C/O Losses—The amount of material lost in a product change. This is the amount of residual material in process lines that must be flushed, any cleaning fluids required, and any off-spec product made getting properties back within spec after the mechanical changeover is complete. In the process industries, material losses often have more of an impact on financial performance than the time lost does. Showing this on the VSM helps clarify what is often the reason for long campaigns.
- Available Time—The total time this process step is scheduled to run. If, for example, a process step is run for two 8-hour shifts for five days per week, the available time would be 80 hours per week. If the equipment is not run during lunch periods and breaks, this might reduce the available time to 70 hours per week. In the process industries, different process steps may have different shift schedules, so available time could vary from step to step.
- Shift Sched—The numbers of hours per shift, shifts per day, days per week, or the total number of hours per day and the number of days per week. For example, 8 × 2 × 5, 12 × 2 × 7, or 24 × 7.

There may be cases where some of this data is not relevant, in which case it wouldn't need to be shown. In other cases, to capture complete understanding of flow, additional items may be listed.

Inventory Data Box

The inventory data box (Figure 2.3) is used for all inventories: raw materials, Work In Process (WIP), and finished product inventory, and should appear wherever that inventory exists in the material flow. Total inventory is shown, without distinguishing safety stock and buffer stock from cycle stock. (These components of inventory are defined in Appendix E and methods for determining those stock requirements are explained.)

Inventory can be shown in the quantity units (pounds, gallons, square feet, etc.) and in the number of containers that are being used. In addition, it should be

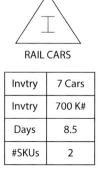

RAIL CARS

Invtry	7 Cars
Invtry	700 K#
Days	8.5
#SKUs	2

FILAMENT TUBS

Invtry	640 Tubs
Invtry	640 K#
Days	9
#SKUs	60

Figure 2.3   Inventory data boxes.

shown in days of supply. This is calculated by dividing the total inventory by the normal throughput, and is equivalent to the average lead time through the inventory. It also shows how long it takes to consume that volume of inventory. Alternately, inventory turns could be shown, but that generally doesn't convey the effect of inventory on flow as clearly. Listing inventory as 9 days gives a more useful picture of the impact on flow than showing it as 40 turns does.

■ Inventory—The average total amount of inventory of all SKUs stored at that position in the process flow. In the process industries, an inventory location may be shared by WIP from several points in the process, so what should be shown here is only the inventory at that stage of production. This should be in units of material volume: pounds, gallons, square feet, rolls, etc.

■ Days of supply—This is the inventory volume converted to a number of days. This can be calculated using a variation of Little's Law:

$$Days\ of\ supply = Inventory/Throughput$$

■ # SKUs—The total number of SKUs or product varieties normally stored at that stage of the process flow. Because many process operations have a high level of product differentiation as material moves through the process steps, this parameter is very important to the understanding of flow dynamics.

Transportation Data Box (Figure 2.4)

For all transportation steps, including deliveries from suppliers of the most significant raw materials and deliveries to warehouses, distribution centers, and customers, the following data should be shown on the VSM:

■ Delivery frequency—How often incoming shipments are received; how often shipments to warehouses and customers are made.

■ Lot size—The average quantity shipped or received. This should be listed in the same units used in order and production data (pounds, gallons, square meters, bales) and perhaps as a number of truckloads or railcars.

■ Transport time—The average time from shipment to receipt.

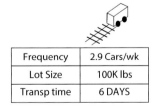

Frequency	2.9 Cars/wk
Lot Size	100K lbs
Transp time	6 DAYS

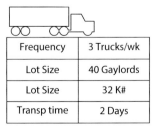

Frequency	3 Trucks/wk
Lot Size	40 Gaylords
Lot Size	32 K#
Transp time	2 Days

Figure 2.4 Transportation data boxes.

Figure 2.5 Customer data box.

Customers Data Box (Figure 2.5)

- Total quantity per unit time—The sum total of all products ordered by all customers, per week, per month, or in whatever time increments the data is collected. The quantity can be expressed in pounds, gallons, square meters, or any other units used by the planning and scheduling processes.
- Takt—Total demand from all customers of this product, reduced to a time base of hours or days. It is the total demand divided by the total time the plant is scheduled to operate during the time increment above. For example, if total customer demand is 100,000 gallons per week, and the plant runs three 8-hour shifts five days per week, Takt is 833 gallons per hour.
- Lead time expectation—The maximum time the customer allows from the time the order is received until material is received by the customer.

If a plant makes two or three distinctly different product types or families, the customers for each product family are often grouped together and shown as a separate customer box for each product family.

Supplier Data Box (Figure 2.6)

The suppliers of key, high-volume raw materials should be shown on the VSM. Not all raw material suppliers should be included; only those who supply a very significant quantity of the total raw material volume normally should be shown. If, however, there is a low volume raw material with lead times or delivery

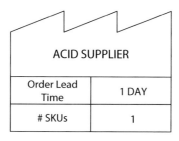

Figure 2.6 Supplier data box.

performance that can likely impact material flow through our manufacturing process, it can be shown to highlight that potential constraint.

■ Order lead time—The time from placing the replenishment order to the time the supplier ships it. The remaining component of total lead time will be listed as transport time on the transportation step. If our supplier is using a make-to-stock (MTS) strategy, the order lead time may be very short—a day or two. If, however, he employs a make-to-order (MTO) strategy, not uncommon in the process industries, order lead time may be several days or weeks.

■ # SKUs—The number of material or part types we normally receive from that supplier is useful information when analyzing our raw material replenishment strategy.

Other relevant information, such as lot size normally ordered, will be listed with the transport data.

Material and Information Flow Icons

Standard icons are used to make the VSM more visual and more readily understood. The icons most commonly used for the material flow and information flow portions of the VSM are shown in Figure 2.7.

Figure 2.8 shows a portion of the material flow on a VSM with the appropriate data boxes and icons.

Information Flow

The top half of a VSM depicts the flow of all information that schedules, manages, and controls the physical material flow. One of the major strengths of the Value Stream Mapping technique is this view of how material flow is enabled or constrained by how the information is being processed. Toyota realized this when developing the "Material and Information Flow Diagram" concept, on which the now commonly accepted VSM format is based.

This connection is important because we have found that in many cases, material flow is limited not by physical bottlenecks inherent in the process equipment, nor in flow problems related to equipment performance, but by mismanagement of demand data, customer orders, and production schedules. When this happens, it creates what are called Capacity Constraints (Umble and Srikanth, 1990), steps in the process that have the inherent capacity to make Takt, but do not because of unsynchronized or inappropriate scheduling. Therefore, showing material flow and information flow on the same map highlights the points of interaction and can focus attention on those that disrupt flow and create waste,

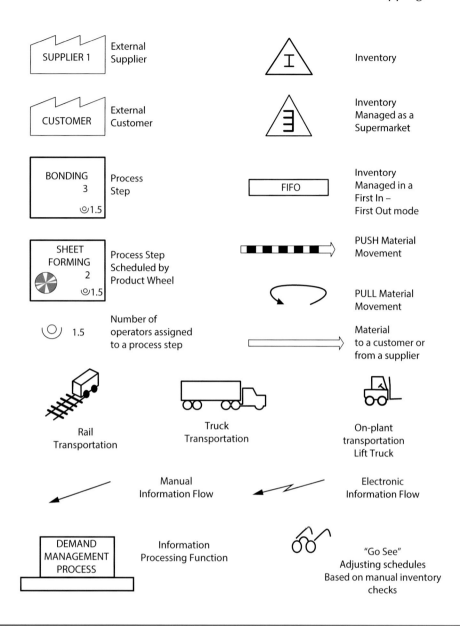

Figure 2.7 Typical VSM icons.

such as breakdowns in communication where people with a role in planning or scheduling processes do not receive needed information in a timely fashion.

Creating the information flow portion of the VSM (Figure 2.9) should start by showing the flow of all incoming customer data, including actual orders and schedules of future demand. The flow of this incoming information is then traced through all of the transactional processes involved in creating daily and longer-term production schedules. Any batching of incoming information or delays in processing should be noted. In electronic ordering processes (Electronic Data Interchange—EDI), information is often processed in real time, but sometimes the systems involved update only once per 24 hours. In more manual processes, orders are sometimes accumulated in weekly batches before processing.

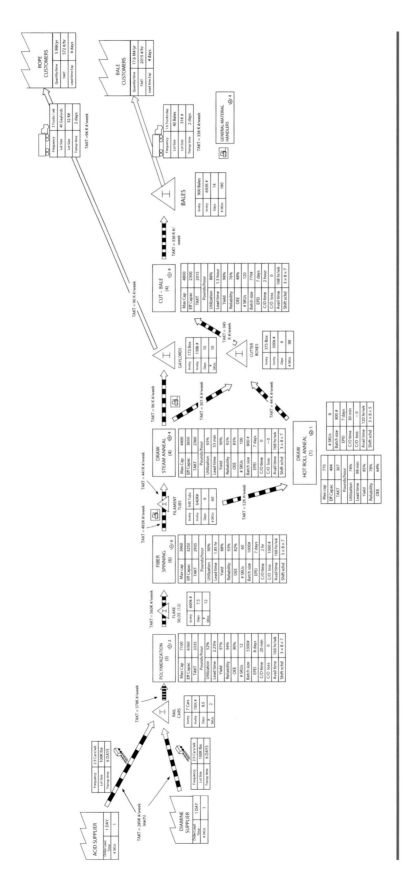

Figure 2.8   VSM material flow with data boxes.

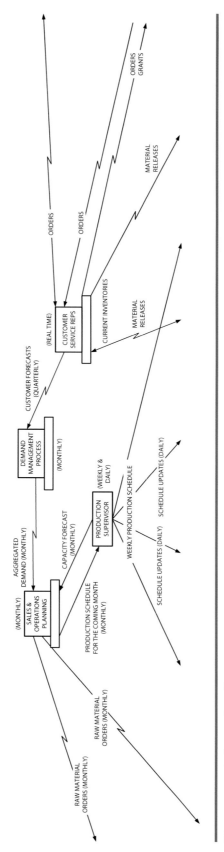

Figure 2.9 Information flow.

In addition, the information processing itself can consume valuable time. Any such delays consume part of the available customer lead time and therefore allow less time for order fulfillment by the manufacturing process. In make-to-stock situations, these delays will require more inventory, and in processes where make-to-order (MTO) or finish-to-order (FTO) would be advantageous, they may even eliminate the potential for MTO or FTO.

The information flow portion of the VSM should show each significant information-processing step as a box, indicating what group performs that step, computer applications such as Material Requirements Planning (MRP), which may be involved, and whether it is a real time, daily, or weekly process. These information boxes are connected by arrows: the usual standard is that zigzag arrows depict electronic information flow and straight arrows depict flow by paper, telephone, or fax. Each arrow should be labeled with a brief (2 to 4 words) description of the content of the information flow.

The flow depicted by these arrows should start with customer input, move through the various information handling processes, and terminate with arrows to the various material flow processes that are scheduled (Figure 2.10). Information arrows should also flow back to suppliers, reflecting how raw material replenishment orders are created and communicated.

The Third VSM Component—The Timeline

The timeline (Figure 2.11) appears as a square wave at the bottom of a VSM, and is intended to contrast non-value-add (NVA) time and value-add (VA) time. In many assembly processes, the NVA time occupies 90% of the total time material is on the plant. In the process industries, this ratio often hits 98%, and can even exceed 99%. It is not unusual for a specific unit of material to be on a plant as raw material, then WIP, and then as finished product for 50 to 60 days, while the time it is being beneficially processed is only a few hours. The purpose of the timeline is to highlight areas where waste adds to overall plant lead time. Many of the forms of waste add time, so the timeline is an indicator of the presence of most waste. Manufacturing Cycle Efficiency (MCE), the VA time divided by the total cycle time, is often used to dramatize this time waste.

The normal convention is that the NVA time is the positive portion, the top of the square wave, while the VA time is the negative or lower part of the wave, although this is not completely standardized across the Lean community; the timeline is occasionally shown with the NVA time as the lower portion of the wave.

It is important to keep in mind that the reason for the timeline is to indicate major areas of time and, therefore, waste. Extreme accuracy is not needed, so you shouldn't spend a lot of time trying to refine the data for the timeline. If the material flow portion of the map has been properly depicted, all of the data necessary to draw the timeline is already included in the data boxes.

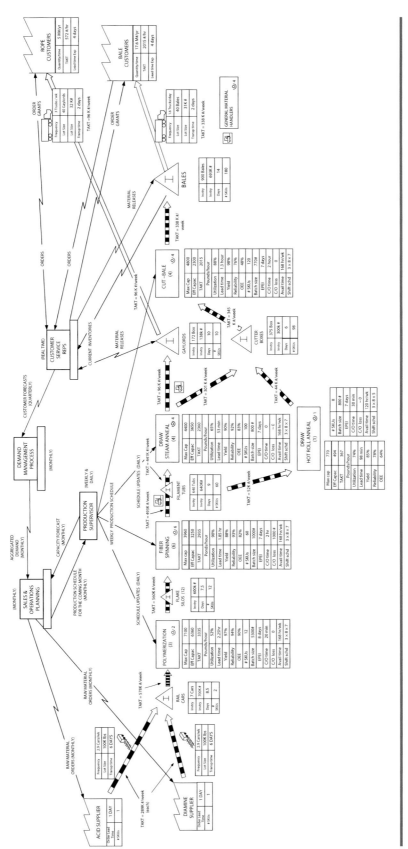

Figure 2.10   Information flow coupled with material flow.

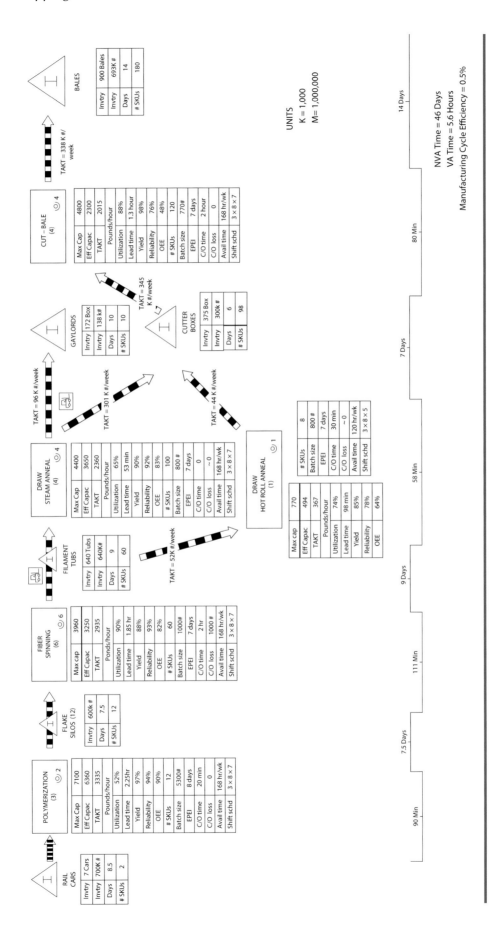

Figure 2.11 The VSM timeline.

When parallel pieces of equipment or process systems have different lead times, VA times, or NVA times, the timeline should be based on a weighted average of the VA time and a weighted average of the NVA time, with the weighting factors based on the proportion of total flow volume through each piece of equipment.

Parallel Equipment

When a step in the process has two or more pieces of equipment (machines, tanks, reactors) or systems (dyeing systems, extrusion systems) in parallel, the recommended practice is to show a single process box and a single data box for that step, with an indication of the number of parallel units. In that case, the data box should show the total maximum capacity, effective capacity, Takt, utilization, and SKU count for the combination all of those pieces.

If the parallel equipment or processing systems have different characteristics that influence material routing or processing in any way, the parallel units should be shown as separate process boxes, with separate data boxes

This is covered in more detail in Chapter 4.

Level of Detail

Keep in mind that a VSM should be a high-level view of a manufacturing process. It should be done in enough detail that flow can be seen and that barriers to smooth flow and process bottlenecks are apparent. It should also have enough detail that the key root causes of poor flow and waste can be diagnosed. However, it should not be so detailed that overall flow cannot be seen and understood. Considering an interstate highway map as an analogy, showing many of the local streets can make it much more difficult to follow the flow of major highways.

Where additional detail is needed, it can be shown on a more specific process map for each of the more complex process steps. These then support the higher level VSM.

Summary

Figure 2.12 shows a complete VSM, with the timeline, information flow, and material flow. The material flow helps us understand all of the material transformation steps, how they interconnect, and how we are creating value for the customer. The information flow illustrates how customer orders, forecasts, and other data at our disposal enable us to manage and schedule the material processing, and the timeline gives an overview of how the manufacturing lead time is being spent and how much of it we are wasting on non-value-adding

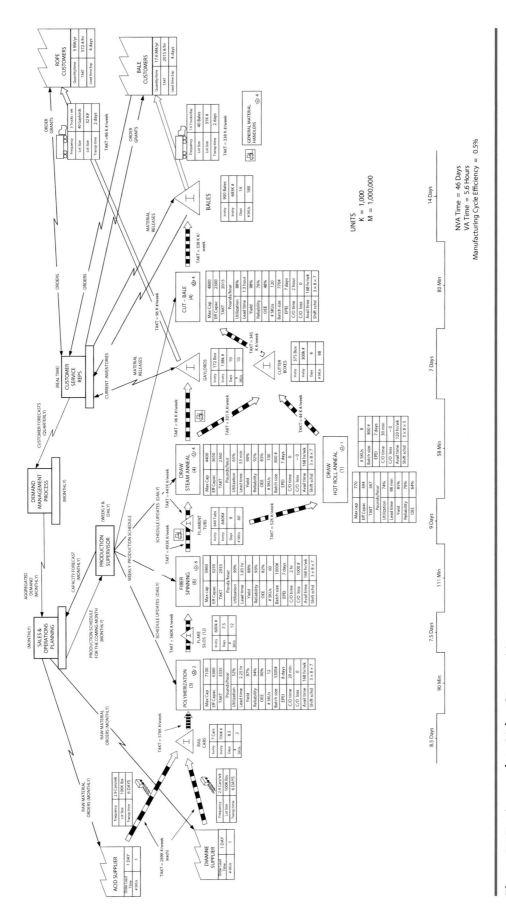

Figure 2.12   A complete Value Stream Map.

activities. With this insight, we can begin to investigate how we can improve the process to improve flow and reduce wasteful activities. That will all be covered in later chapters, but first, now that we understand the structure and components of traditional VSMs, we'll shift our focus to the adjustments that should be made for process operations.

Chapter 3

VSM Enhancements for Process Operations

Distinguishing Features of Process Operations That Require a Different VSM Approach

There are a number of characteristics of process operations that distinguish them from discrete parts assembly operations, characteristics that require a somewhat different approach to the applications of Lean principles, as described in *Lean for the Process Industries—Dealing With Complexity*. In this chapter, we'll focus on those with a significant impact on how a VSM should be drawn. Appendix A highlights some of the others.

Capital Intensive vs. Labor Intensive

One of the more obvious differences is that process operations tend to be very capital intensive, where assembly processes tend to be very labor intensive. Process plants often have very large, very expensive equipment, such as large mixing tanks and reaction vessels with thousands of feet of process piping, or large machinery capable of casting very wide plastic or paper sheets or extruding plastic pellets. The tanks and piping may be jacketed with heating media like steam or Dowtherm, adding to the size, weight, and investment. In contrast, much of the machinery found in discrete parts plants is relatively small with simple electrical and compressed air connections. While far less capital intense, a parts manufacturing plant is generally much more labor intense. A discrete parts plant is likely to have a very high labor force, while a very large chemical or food plant may have less than 100 operators to staff a 24-hour per day operation.

This is relevant because traditional Lean is focused on eliminating wasted labor and improving labor productivity, so a VSM will show the number of operators assigned to each step in the operation and to auxiliary tasks like driving lift trucks. Labor productivity is also an important factor in the process industries, but asset productivity is typically far more important, so a process VSM must highlight asset performance factors and associated wastes. Wasted time on the production equipment generally has far more negative impact than wasted labor.

A consequence of the asset intensity is that throughput is limited by equipment rather than labor. In many assembly processes, bottlenecks can be eliminated and throughput increased by adding people. This is rarely the case in the process industries. Applying more labor to a paper sheet forming machine or a batch paint-mixing vessel will not increase throughput one iota. In addition, because many process plants run around the clock, seven days per week, adding extra shifts to open bottlenecks is not an option. So the understanding of root causes of waste, and changes necessary to reduce it, must focus much more on the process equipment than on how labor is applied.

A traditional VSM typically lacks the breakdown of factors detracting from asset productivity, such as Overall Equipment Effectiveness (OEE) and its components: yield, operating rate, and equipment reliability. These are critical to understanding waste and its root causes in process operations. Another parameter not normally shown on a traditional VSM is equipment utilization, demand divided by capacity, a key to understanding bottlenecks and flow limitations.

Therefore, a process plant VSM must include all of this information on equipment performance to allow a complete understanding of the wastes affecting throughput and to understand what must be done to eliminate them.

Material Flow Patterns—SKU Fan Out

Perhaps the most dramatic difference between assembly plants and process industry plants is that the flow patterns and flow dynamics are quite different; in fact, they are the opposite of each other. As described by Umble and Srikanth in *Synchronous Manufacturing* (1990), the predominant flow characteristic in an assembly plant is *convergence* of part types, whereas process plants are characterized by product type *divergence*. This difference is shown schematically in Figures 3.1 and 3.2. Figure 3.1 represents a typical assembly plant, with material flow from bottom to top. At the start, there may be hundreds, thousands, or tens of thousands of individual screws, nuts, bolts, springs, sheets of plastic and sheet metal, etc. As these parts are processed and assembled into sub-assemblies, then sub-systems, then complete systems, and finally to the finished product, the number of different part types diminishes dramatically. There is a significant convergence of parts as we move toward the final product. Considering the manufacture of a Toyota Camry, for example, there may be tens of thousands

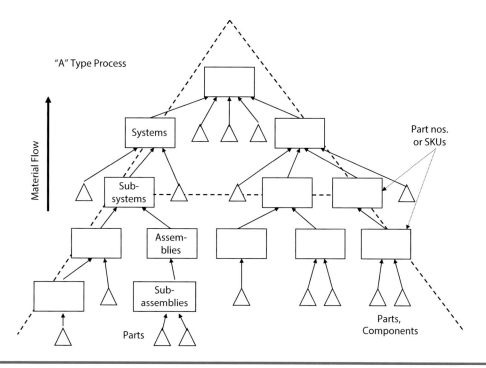

Figure 3.1   Schematic diagram of an "A" type process.

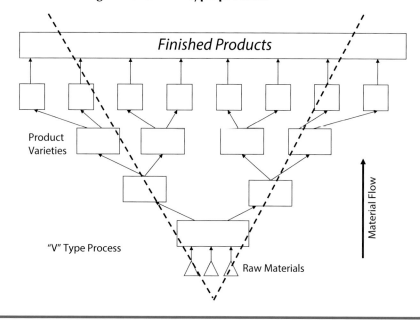

Figure 3.2   Schematic diagram of a "V" type process.

of part types at the start, while the final product comprises two trim lines, in several colors, for a total end item variety of perhaps a dozen or so. This flow pattern has been called an "A" type process, because it resembles the letter A when diagrammed as in Figure 3.1.

Process industry operations generally follow the flow pattern shown in Figure 3.2. The process starts with very few raw materials, which may be mixed, reacted, then cast or extruded as fibers, sheets, or pellets, and then further processed to create tremendous final product variety. In the manufacture of nylon yarns for apparel or for seat belts, the process starts with very few raw materials: adipic acid, diamine, TiO$_2$, and demineralized water. These are mixed, polymerized to create a highly viscous molten plastic, and then extruded as groups of very fine fibers. They can then be stretched at different ratios to build a desired tenacity level, annealed to permanently set the properties, dyed to fit the particular end use, and wound on one of a wide variety of rolls or spools. What started as four primary ingredients ends as hundreds or thousands of end item SKUs. The predominant flow pattern is one of divergence. This flow pattern is sometimes called a "V" type process, reflecting the way it is diagrammed in Figure 3.2.

As a specific example of a "V" type process, the product type fan out for the fiber spinning, stretching, cutting, and baling plant we will be referencing throughout this book is shown in Figure 3.3. Two raw materials, adipic acid and hexamethylene diamine, are stored in input tanks. They are then mixed with de-ionized water and reacted at various temperatures and pressures to form 12 types of molten polymer, which are extruded and chopped into pellets called flake and stored in large silos. The flake is later pneumatically conveyed to a fiber-spinning machine, which melts the flake and extrudes it as very fine fibers; a fiber bundle may be comprised of dozens or hundreds of extremely fine filaments. A number of process variables in spinning can generate 60 types of fiber. The spun filaments are stored in large stainless steel tubs, and then sent to machines that stretch ("draw") and heat set ("anneal") the fibers. Various combinations of drawing and annealing conditions can create 108 types of fiber. Of those 108 types, 10 are sold as is, and 98 are cut into very short lengths and

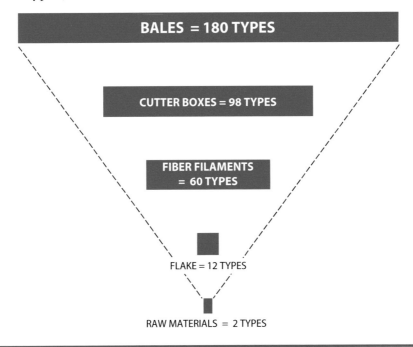

Figure 3.3 Example of a "V" type process—manufacture of cut fibers.

baled, in a process similar to the way cotton fibers are baled. Various cutting and baling settings can create 180 types of baled fiber.

So in this process flow, we have significant product differentiation at each major step: at polymerization (2 → 12), at fiber spinning (12 → 60), at draw–anneal (60 → 108), and at cut–bale (98 → 180). We have transformed two types of raw materials into 190 final products (10 sold as uncut fiber and 180 cut and baled). This high level of divergence, of product differentiation, is very typical of the "V" processes found in process plants.

The same general product fan out is found in beverage manufacture, where sugar, water, and a few flavorings can result in hundreds of cola products, with regular and diet, caffeinated and caffeine free, and various fruit flavored varieties, sold in all combinations of these and in many different can and bottle sizes.

The presence of a number of product differentiation points, where a single material can be transformed into one of several varieties, is at the core of the difference in A type and V type processes. Scheduling is critical at each differentiation point; if the wrong differentiation decisions are made, it can have a profound impact on manufacturing system performance. If differentiation decisions are made incorrectly, if an unneeded variety is produced, it flows to finished product inventory, filling the warehouse with currently unneeded product. It has also consumed valuable production capacity, which is then unavailable to make needed products. Therefore, finished product inventory can be very large, but with poor customer service. With the previous fiber example, if the product is cut to currently unneeded lengths, finished product inventory of those cut lengths will increase while shortages of others can occur.

This diverging flow pattern has two implications for a process VSM:

1. The information flow portion of the VSM should indicate whether these differentiating decisions are being made in a timely fashion, based on current inventory and demand status.
2. Because this product type fan out has such an influence on flow dynamics, showing the number of product types, or SKUs, created at each step is essential to a process VSM.

Product Changeover Issues Are Complex

In discrete parts manufacturing, product changes often involve changing tooling and then adjusting or calibrating the machine with the new tooling. The primary wastes are time and labor. In the process industries, there can be machine changes to be made, but these are often a small component of the total waste. In addition to lost productivity and wasted labor, there can be lost process materials and cleaning fluids and solvents, and requirements for additional test laboratory time to ensure that the new product type is within specifications. For example, in the food industry, product families often have products that contain allergens such as peanuts or dairy ingredients, and those that do not. Where volume warrants it, dedicated lines are often the answer, but in many cases, this is not economically justified. Consequently, very extensive clean-ups must be done after running the allergen-based products, with complex decontamination processes, and testing to ensure a contamination-free environment afterward. In the synthetic rubber industry, process capability is sometimes poor enough to require significant production time to get properties such as viscosity back within specifications after a changeover, thus generating substantial amounts of wasted material. Again, significant testing laboratory time and facilities can be required to determine when properties are finally on aim. These factors drive production schedulers toward longer campaigns.

In the chemical industry, where many processes run at high temperature and pressure, a lot of time can be lost waiting for the process equipment to reach new temperature, pressure, and other parameter settings after the mechanical changes are complete. This also tends to encourage a very large campaign to be run on the current material prior to switching to the next. This, of course, is overproduction, and results in high finished product inventory as well as constraining flow of other products.

Because changeover times and material losses can have such an influence on campaign lengths and therefore inventory waste, both should be shown in the VSM process data boxes. It is also necessary to include the overall campaign cycle time, the EPEI (Every Part Every Interval) in Lean terms, to see the negative influence of long or costly changeovers on campaign lengths and inventory.

Product Families—Selecting a Target Product or Family

Selecting a subset of the full range of products as the basis for the VSM can simplify both the flow mapping and the data gathering effort and is often recommended by VSM guides. However, focusing on a single product family can be misleading and understate material flow in a process industry plant. In many cases, all product families flow through the same assets, so including only a subset of the full flow can create a misleading view of how fully each asset is utilized and hide bottlenecks. So where assets are shared across all products, the VSM should be based on the full flow, not a single family. Admittedly, this can make the VSM more complex, but this degree of complexity is necessary to get a clear, holistic view of flow.

Takt Rate vs. Takt Time

In the last chapter, we mentioned that in traditional Lean, Takt is a time-based parameter, while we find that it is clearer and more useful as a rate-based parameter in most process operations.

As an example of its use in traditional Lean, a lawn mower producer has demand of 2000 mowers per week, and runs the plant on two 8-hour shifts for five days per week. The available time is 80 hours per week, so Takt is 80 hours divided by 2000 mowers, or 2.4 minutes. That is, the plant must produce a mower every 2.4 minutes to meet customer demand. If the manufacturer can synchronize all operations to the 2.4-minute Takt, all customer demand can be met without any overproduction. In addition, if every mower has 4 wheels, the Takt for the wheel production is 0.6 minutes, that is, a wheel must be produced every 0.6 minutes to meet customer demand.

That's a reasonable way to look at production needs in a parts manufacturing operation, but far less so in a process operation. A good example is a salad dressing line, with a rotary bottle filler capable of 300 BPM (bottles per minute), and a Takt rate of 270 BPM. Those numbers have relevance for the people running the line, and they can easily see that the filler must run 90% of the time to meet customer demand. But those same numbers give a Takt time of 0.22 seconds per bottle and a cycle time (the time-based analog of effective capacity) of 0.2 seconds, which have far less meaning to those managing and operating the line. If sales growth raises Takt to 310 BPM and it becomes necessary to raise effective capacity to 340 BPM, mechanics and engineers can judge the practicality of doing so. However, if they are told that cycle time must be reduced from 0.2 seconds per bottle to 0.176 seconds, the numbers don't have nearly as much relevance.

Because one of the key purposes of a VSM is to present a clearly understood picture of flow and the influencing factors, the governing principle should be to present information in the most relevant manner. For these reasons, Takt and capacity should be expressed either as time parameters or as rate parameters based on what the operations personnel are more familiar with and can easily comprehend. The two are mathematically equivalent, the rate values being simply the reciprocal of the time parameters. Whatever the choice may be, it should be followed consistently throughout the map.

Units of Production

One of the more important decisions you have to make when starting the material flow portion of the map is which units of production to use for each process step. In a process plant, the units we use to measure throughput can be very different from step to step, and the choice of units must be made thoughtfully.

In a ketchup bottling plant, for example, the raw materials include sugar, bought by the pound, tomato puree, bought by the pound or perhaps by the crate, vinegar by the gallon, and salt and spices by the pound. The ketchup coming from the kitchen will be measured in gallons. The output of the bottle-filling machine will be measured in bottles, the case packer in cases, and the palletizer in cases or pallets. The numerical relationship between these parameters will change with product type: the number of bottles per case will be greater with small bottle sizes, and the number of ounces per bottle will vary with bottle size. This is quite a contrast with a lawnmower factory, for instance, where the number of wheels per mower is always four, regardless of the blade or engine size. Moreover, there will always be one carburetor per mower, completely independent of the engine horsepower or any other factor.

There is sometimes a desire to use the same units throughout the map, as this may appear to simplify things. On the contrary, this usually leads to confusion and errors. In the ketchup plant, if we were to use bottles as the throughput parameter for the entire VSM, then the throughput from the kitchen could not be determined because it depends on the product bottle size mix. Likewise, we couldn't use ounces or gallons because the bottle filler throughput depends on the bottle size being filled. Therefore, the units used for Takt and for capacity must be different at different steps in the process, and the choice, while very important, is not always obvious.

The best answer to this question is to use units related to the rate determining parameter for the equipment at each step. In a plant making paper for computer printers, books, and magazines, the rate determining parameter for the sheet forming system may be the linear speed of the apparatus winding up the formed rolls; the system may be capable of producing just as many 12-foot wide rolls as it is of producing 10-foot wide rolls, even though the square footage is 20% greater. Therefore, for the roll forming system, linear feet may be the most appropriate unit; if all master rolls were the same length, rolls would also be an appropriate unit. So in that case, the effective capacity could be stated in terms of master rolls per hour, which would be constant regardless of the specific product being formed. If master rolls have different lengths depending on product type, then linear feet would be the more appropriate parameter.

Takt and capacity can be defined in input or output quantities for any process step, whichever is most directly related to throughput capability. In the paper making plant, in the step that slits the wide master rolls into narrow rolls, throughput is related to the number of incoming rolls to be slit, not slit rolls created by that step. It takes the same time to slit an incoming roll of a given size into two slit rolls as it does to slit it into six slit rolls. Therefore, if Takt and capacity are stated in units of incoming master rolls, those values will be constant regardless of the product type and therefore the number of slits. Thus, in that situation, the input material is the rate determining parameter. A carton packing operation, where several bottles are packed in each carton, may have its throughput governed more by the number of completed cartons than by the number of items in each carton, so in that case we would use the output parameter. The rate of a paint container filling line may be more related to the capacity of the pumping system than it is to the number of containers (pails, drums, etc.) being filled. A 1000-gallon tank usually takes approximately the same amount of time to pump out, regardless of whether it is being pumped into 55-gallon drums or 5-gallon pails, so the Takt for a filling station would be based on the inherent capacity of the system, rather than the number of output drums or pails.

The governing principle should be to define Takt and capacity in whatever units reflect the throughput capability of the equipment which is most consistent across all products. Of course, for any step in the process, the same units must be used for Takt and for effective capacity at that specific step.

Generating the Map

In *Learning to See*, Rother and Shook recommend that the VSM be created by walking the line and sketching it on a pad as you go. This is often difficult in a process plant because the physical equipment arrangement does not mirror the logical flow pattern. The equipment may be somewhat scattered over large areas of the plant floor or across several buildings, so it can become difficult to follow the process flow without having a clear understanding of flow beforehand. It is also sometimes true that no single individual understands process flow in detail, so the knowledge required to create the VSM necessitates getting a team together. This is not intended to diminish the value of walking the process! On the contrary, there is significant value in observing the process and interacting with the people on the floor, which should not be understated. Go to the Gemba, to the place where the activity occurs, is a key Lean principle that is as vital to process operations as it is to assembly manufacturing. As stated so eloquently by Yogi Berra, "You can observe a lot by just watching." However, it must be recognized that for many process industry lines, it is not sufficient to "see" the process. In some cases, it is better to create the skeleton of the material flow portion of the map before a thorough process walk is taken, so that you have some context for what you are seeing.

Our typical practice is to form a team of operators, mechanics, process engineers, and perhaps QC lab operators and shipping clerks, and to convene in a conference room. We then create a draft of the material flow portion of the map, and get it printed at a size that is convenient to take out on the plant floor and mark on. We then have an understanding of overall process flow and can begin to walk the process with a better idea of what we are seeing and how it all fits together, and record more details of what we see.

It is often suggested that the VSM be drawn by hand before capturing it electronically. We have found that creating it electronically at the start can be very effective. We'll have a computer running a mapping/charting application like Microsoft Visio or iGrafx Professional, with the map projected onto the conference room wall, so everyone can see the additions and corrections as they are being made. Where hand-drawn charts can get very sloppy and hard to read if very many changes are made, the computer-generated picture is always clean. These tools make it very easy to open up space if an additional step must be added, and to rearrange process boxes to maintain a smooth view of flow.

Experience has taught us that because the map is easy to modify and at the same time keep legible, participants are less reticent about correcting errors or making necessary additions. The electronic version becomes the permanent (permanent, that is, until the next change is made) record of the map and facilitates the printing of copies to be distributed to others for comment and upgrade. Some plant teams favor the hand-drawn "paper on the wall" approach and that also works very well. It has the advantage that it makes participants more physically involved in the mapping process. It should be left up to the mapping team to make the choice on a computer keyboard or Sharpie pens based on what they are most comfortable with.

Time Units

Some VSM books recommend that seconds be used as the time unit. In process operations, minutes are generally more meaningful, and in some cases hours is the most practical unit.

Where to Begin

Several VSM guides recommend starting at the shipping dock and working backward through the process to develop the map, to emphasize that all production activities should start with the customer and what he or she needs. For converging manufacturing processes ("A" type processes) this makes sense, is effective, and has the benefit of placing the initial focus on the customer. However, in diverging ("V" type) processes, with a very high level of differentiation and therefore a high number of end products, it can sometimes be confusing to start at the final products. In those cases, it is easier to see flow, to describe and discuss it, by starting at the less complex front end, where few raw materials are introduced. It is often easier to trace flow by following the diverging paths of material as it flows through the process toward the customer, so that is generally the practice for process VSMs. It is up to the mapping team to decide which makes more sense in their specific situation, and their responsibility to make sure that each step is described with the final customer in mind.

Summary

The way you make salad dressing, house paint, shampoo, and automotive antifreeze is dramatically different from the way you make lawn mowers or cell phones. Therefore, the VSM describing those operations must also be different. While the basic structure of the map is the same as that used for traditional VSMs as described in Chapter 2, additional data must be included to fully characterize process operations, and some of it must be presented in a more relevant manner.

The fact that equipment is most often the rate-limiting factor means that the VSM must include more detail on equipment performance and OEE factors. To fully understand the degree of material type divergence, the SKU count at the exit of each process step should be shown, with the information flow done in enough detail to understand what factors are included in making differentiation decisions. Because changeovers can waste much more than time, material losses should be listed in the data boxes.

Rather than basing the map on a pilot product family, all products must be included to avoid understating utilization and the potential for bottlenecks. In addition, showing Takt and capacity as rate values rather than time values usually makes the numbers more relevant and more useful to those familiar with the process.

In later chapters, we'll see how these factors apply as we draw a VSM for an actual process operation, but first we will review a few additional best practices that apply to any VSM whether it is for a process operation or parts assembly manufacturing.

Chapter 4

Additional Good Mapping Practices

Good VSM Practices

Now that we have covered the unique features that must be included on a VSM to thoroughly describe a process operation, some general practices that apply to any type of VSM will be described.

Map Layout—Flow Direction

VSMs should always depict flow from left to right in one relatively straight line. Because of the complexity of some processes, there can be a tendency to locate process steps on the map such that flow lines go up, down, or right to left in an attempt to fit everything in and keep the overall map within reasonable dimensions. This usually obscures a clear view of flow. When this occurs, it is usually because too much detail is being included, that is, the process is broken down into too many steps. When that happens, you should step back and decide if some process steps should be grouped together into a single combined step. If it is necessary to show those processes as separate steps to clearly depict flow, then having a wider map is acceptable and the right thing to do. It is far more important that the VSM create a clear and accurate picture of flow and its barriers and wastes than that it fit within a certain size.

The point is that if a high level of detail is required, you should avoid the temptation to use vertical space on the map to keep it narrow enough to fit some overall dimensional goal.

Level of Detail

That brings us to our next point—you should decide very carefully how much detail is needed to satisfy the objectives for which the VSM is being created. With too much detail, the overall flow dynamics can be buried in the important but not critical details.

In a carpet manufacturing facility, the dyeing step consists of a complex array of mix tanks, heaters, a trough called a dye bath through which the carpet is moved, and then a drying step. In order to provide the right balance of complexity vs. flow visibility, this could all be shown as one step called "dyeing." Most of the parameters in the data box, especially Takt, effective capacity, yield, changeover time, and changeover loss, are far more relevant to the overall dyeing process than to its individual components.

If additional detail is needed to better understand specific wastes involved in dyeing and their root causes, it can be shown on another map specific to the dyeing step. It is often the case that lower level maps are needed to show details not included on the higher level VSM. In some cases, these lower level maps can follow the VSM format, while in other cases more traditional process mapping is appropriate.

This is sort of a "divide and conquer" approach to understanding overall flow and its detractors, and also understanding enough of the details of those detractors to know what must be done to resolve them.

Level of Precision

You shouldn't spend a lot of energy trying to refine the precision of the data on the map. The numbers are on the map to give clues where the wastes are, and where flow is not synchronized with Takt. Close approximations are good enough to satisfy this purpose. You may notice that some rounding is done in the data box calculations in Chapters 7, 8, and 9, and that's perfectly acceptable. In some cases, the source numbers used to calculate map parameters are not perfect; so, good judgment is required to determine if they are good enough.

Parallel Equipment

In the flow through many process plants there are several machines or vessels in parallel at a step, each of which can perform essentially the same process operation. This is generally because the total flow requirements exceed the capacity of a single piece of equipment, or because different products have different requirements necessitating specialized equipment. If the parallel equipment is identical, or at least similar in all relevant capabilities, it should be shown on the VSM as one process box, with a notation that there are N units in parallel. In that case, the data box should show the total maximum capacity, effective capacity, Takt, utilization, and SKU count for the combination of all of those pieces. If the reliability, yield, or changeover time differ slightly, the weighted average should be shown. If they differ significantly, they probably should be shown as separate process boxes with separate data boxes. In addition, if the parallel equipment or processing systems have any other differing characteristics that influence material routing or processing in any way, the parallel units should be shown as separate process boxes, with separate data boxes.

In some situations, parallel equipment is designed for very different throughput capacity, where the higher speed equipment is reserved for high volume products and the slower equipment for the products with lower market demand. Different pieces of parallel equipment may have different reliability history, different yields, or even be staffed on different schedules (very specialized equipment needed for only a small portion of the overall flow may need to be staffed only eight hours per day, for example). The data boxes should reflect all of these differences.

In the fiber spinning and cutting process we will describe in detail in the next chapter, five machines do the stretching ("drawing") and heat setting ("annealing") processes. One of these is a very old machine that uses hot rolls to accomplish the annealing after the drawing step. Four are relatively modern machines that use steam to heat the fiber for annealing. The older machine has significantly different operating characteristics: poorer reliability, lower yield, and longer changeovers. So it is shown as a separate process box and data box, as shown in Figure 4.1.

Another consideration is that there are a few products that must be run on the older hot roll annealer. It is likely that they could be run on the steam anneal machines, but the specific process conditions for that haven't been determined, and the product end-use qualification tests have not been run. Therefore, for now, they must be run on the hot roll machine, and that's another reason to show this as a separate process box, to reflect the Takt of those products vs. the Takt of the products run on the steam annealers.

The process step following the draw–anneal step is the cutting of the long fiber rope into very short (1/2 in. to 1½ in.) lengths and then baling them using the same type of machinery that is used to bale cotton. There are four cutter–balers in parallel, and while there are some performance differences between them, none is significant enough to change the data to be put into the data boxes. For example, the reliability factors of the four machines range from 73% to 78%, but showing the average of 76% is accurate enough for a VSM at this level, so the four are shown as one process box, as shown in Figure 4.2.

Logical Flow vs. Geographic Arrangement

For a VSM to provide the appropriate insight into material flow, and into any detractors to smooth, synchronized flow, it is vital that flow be shown on a logical rather than on a geographically correct basis.

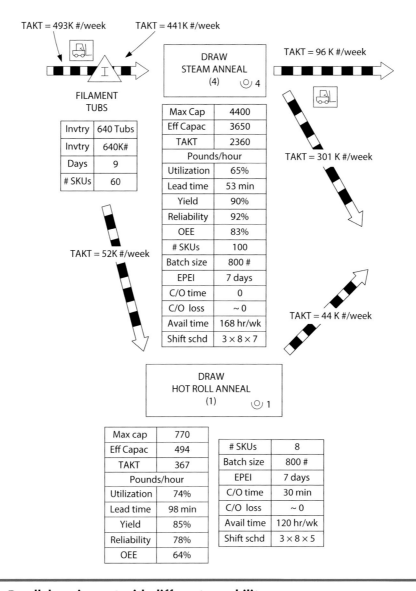

TAKT = 493K #/week   TAKT = 441K #/week

TAKT = 96 K #/week

TAKT = 301 K #/week

TAKT = 52K #/week

TAKT = 44 K #/week

FILAMENT TUBS

Invtry	640 Tubs
Invtry	640K#
Days	9
# SKUs	60

DRAW STEAM ANNEAL (4)   4

Max Cap	4400
Eff Capac	3650
TAKT	2360
Pounds/hour	
Utilization	65%
Lead time	53 min
Yield	90%
Reliability	92%
OEE	83%
# SKUs	100
Batch size	800 #
EPEI	7 days
C/O time	0
C/O loss	~ 0
Avail time	168 hr/wk
Shift schd	3 × 8 × 7

DRAW HOT ROLL ANNEAL (1)   1

Max cap	770		# SKUs	8
Eff Capac	494		Batch size	800 #
TAKT	367		EPEI	7 days
Pounds/hour			C/O time	30 min
Utilization	74%		C/O loss	~ 0
Lead time	98 min		Avail time	120 hr/wk
Yield	85%		Shift schd	3 × 8 × 5
Reliability	78%			
OEE	64%			

Figure 4.1 Parallel equipment with different capability.

There is sometimes a tendency to arrange the process boxes on the VSM in approximately the same relationship as their physical arrangement on the plant floor. This can have a useful purpose in highlighting the transportation waste and its influence on transportation lot size and therefore production lot size. It may also have a benefit in understanding opportunities to relocate equipment to reduce transportation waste, although equipment relocation is usually economically impractical in the process industries due to equipment size and connections.

However, any benefit gained by depicting a geographically accurate layout is minor compared to the loss of a clear view of logical flow, which is vital to our understanding of the transformations being made to material to meet customer requirements, and the wasteful things being done along the way.

As an example, consider a sheet goods process making paper for computer printers. The process consists of a sheet-casting step, a heat setting or annealing

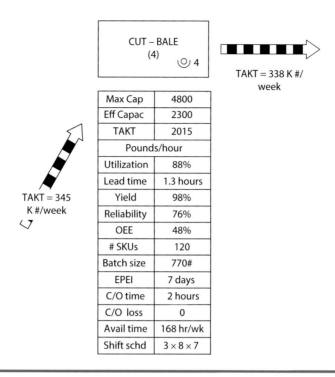

CUT – BALE (4) ⊙ 4	
Max Cap	4800
Eff Capac	2300
TAKT	2015
Pounds/hour	
Utilization	88%
Lead time	1.3 hours
Yield	98%
Reliability	76%
OEE	48%
# SKUs	120
Batch size	770#
EPEI	7 days
C/O time	2 hours
C/O loss	0
Avail time	168 hr/wk
Shift schd	3 × 8 × 7

TAKT = 338 K #/ week

TAKT = 345 K #/week

Figure 4.2 Parallel equipment with similar capability.

step, a slitting step that cuts the very wide cast rolls into narrower widths required by end uses, and a cutting step that cuts across the roll to form individual sheets. There is significant work in process between several of these steps, which is stored in a centrally located Automatic Storage and Retrieval System (ASRS), a high-rise rack system using automatically guided cranes to transport a roll to the designated slot. A VSM based on the geographic arrangement would look like that shown in Figure 4.3.

Although geographically accurate, this view provides no sense of material flow or the sequence in which specific operations are performed.

Figure 4.4 shows a much better view of this situation. The three inventories shown are actually at the same location, but have different functions, and store material in different forms at different points in the process flow, which is clearly depicted in this representation.

The fundamental principle is that the VSM should be arranged so that true flow can be seen. If storage areas and testing labs must be shown several times on the map to accomplish this, it should be done. There is often a need to see flow on a geographic basis, especially where some equipment relocation may be possible, and that should then be depicted on a separate plant layout or floor plan.

Support Processes

The question is often asked if support facilities such as quality testing labs should be shown on the VSM. The general answer is that this is not typically done.

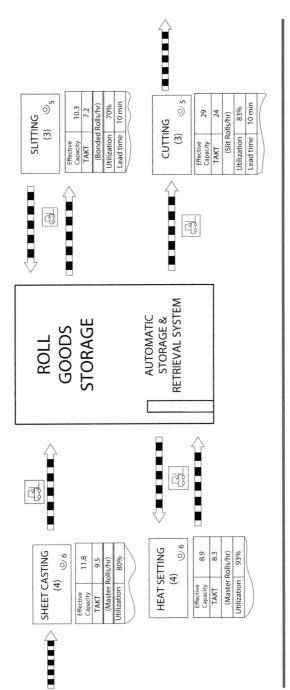

Figure 4.3 Geographic representation on a VSM.

SHEET CASTING (4) ⏱ 6

Effective Capacity	11.8
TAKT	9.5
	(Master Rolls/hr)
Utilization	80%
Lead time	15 min
Yield	87%
Reliability	90%
UPtime	73.6%
# SKUs	50
Batch size	1 roll
EPEI	7 days
C/O time	1 hr
C/O loss	2 Rolls
Avail time	168 hr/wk
Shift schd	3 × 8 × 7

Invtry	1990	Rolls
Days	10	
# SKUs	50	

HEAT SETTING (4) ⏱ 6

Effective Capacity	8.9
TAKT	8.3
	(Master Rolls/hr)
Utilization	93%
Lead time	17 min
Yield	87%
Reliability	98%
UPtime	61%
# SKUs	200
Batch size	1 roll
EPEI	7 days
C/O time	45 min
C/O loss	~ 0
Avail time	168 hr/wk
Shift schd	3 × 8 × 7

Invtry	1380	Rolls
Days	8	
# SKUs	200	

SLITTING (3) ⏱ 5

Effective Capacity	10.3
TAKT	7.2
	(Bonded Rolls/hr)
Utilization	70%
Lead time	10 min
Yield	98%
Reliability	95%
UPtime	69%
# SKUs	1000
Batch size	1 roll
EPEI	7 days
C/O time	5 min
C/O loss	~ 0
Avail time	168 hr/wk
Shift schd	3 × 8 × 7

Invtry	4000	
Days	7	
# SKUs	1000	

CUTTING (3) ⏱ 5

Effective Capacity	29
TAKT	24
	(Slit Rolls/hr)
Utilization	83%
Lead time	10 min
Yield	100%
Reliability	98%
UPtime	98%
# SKUs	1800
Batch size	1 Slit Roll
EPEI	7 days
C/O time	0
C/O loss	0
Avail time	168 hr/wk
Shift schd	3 × 8 × 7

WRAPPING PACKAGING LABELING ⏱ 4

Effective Capacity	200
TAKT	120
	(Cut Rolls/hr)
Utilization	60%
Lead time	8 min
Yield	100%
Reliability	98%
UPtime	98%
# SKUs	2000
Batch size	1 Cut Roll
EPEI	7 days
C/O time	0
C/O loss	0
Avail time	168 hr/wk
Shift schd	3 × 8 × 7

Figure 4.4   Logical representation on a VSM.

If, however, the support facility introduces delays in primary material flow or impacts it in any way, then it is a candidate to be shown. For example, in some processes, material must be tested for quality parameters before it can be released to the next step in the process, and the test time inserts a delay into the process. In that situation, it is very appropriate to show the testing step on the map, especially as it affects the VSM timeline. The testing step should include a data box that has enough information to clearly show the impact it has on flow.

Similarly, if limited supply of utilities such as steam or compressed air can limit process flow, it must be reflected on the map, either as a separate process box or as a note under the data box for the affected steps.

Computer Tools vs. Brown Paper

There is an ongoing debate about whether it is better to create the first version of the VSM on a large sheet of brown construction paper, using Post-it® notes for the process steps and data boxes, or to enter it directly into a computer tool like Microsoft Visio. Traditionalists favor the brown paper approach, and it has advantages. It gets the team more physically involved in the map drawing activity, and creates a more "this is real" aura, where a map on a computer screen may seem more abstract.

I have found that the use of a computer can be very effective. I often create the first version of the VSM using Visio or iGrafx, with the image being displayed on a large screen for the entire team to see. Although the team is seated, not standing, and therefore less physically involved, they are certainly as emotionally and intellectually engaged and generally display a high level of involvement and enthusiasm for the effort. A key advantage of this approach is that it is far easier to correct mistakes and still keep the map clean. If the team decides to add a step inbetween two currently shown steps, it is very easy to open up the space to do that rather than having to crowd it in. In addition, because it is so simple to do, there is less reluctance to make changes for fear of messing things up. Another advantage is that the current version of the map can be printed and given to team members to take out on the floor to validate the paths or data shown. And it can be e-mailed to others to get their input.

In the end, the choice should be based on whatever the team is most comfortable with. But the computer approach should not be ruled out simply because of an adherence to tradition. My personal preference is to use brown paper when I'm teaching VSM classes, as that gets me up in front of the class and has them focused on me and what I'm saying as well as on the map. Nevertheless, I generally prefer the electronic method when I'm working with a team on a real application, but am happy to use the brown paper approach if that's what the team prefers.

If a computer is to be used, either for the first draft or for the final version, there are a number of tools available: Visio, iGrafx Process, eVSM, LeanView, MiniTab, and others. Some of these go well beyond the basic charting capability, and can perform calculations on the data in the data boxes. However, as of now, they are designed around traditional Lean VSMs, for discrete parts flow and assembly, and have strong limitations when it comes to the process-specific VSM features we discussed in the last chapter. Nevertheless, this is a continually evolving field, so it warrants a close look at current offerings and capability when you begin your VSM activity.

Qualified Guidance and Coaching

The team creating the VSM should include representatives of all the groups or functions that participate in executing the process, including operators, mechanics, lab technicians, quality associates, process engineers, planners, and schedulers. It will be difficult to create a complete map without their input, or to move toward an improved future state without their participation in designing it.

It is also critical that the team has someone with strong Lean experience:

■ To provide guidance in the creation of the VSM
■ To ensure that the team understands the full potential of Lean tools like SMED, cellular flow, 5S, visual management, and pull replenishment
■ To ensure that all improvements are being scoped based on sound Lean principles
■ To provide whatever teaching and coaching that may be required

A lot of Lean, and even Value Stream Mapping, can appear deceptively simple, enough so that it is sometimes attempted by highly intelligent, well-intentioned novices. This usually leads to less than satisfactory results. As Jeffrey Liker (*The Toyota Way Fieldbook*, 2006) points out, "Mapping makes people feel like they're doing Lean, but it is simply drawing pictures…if I hand you a blueprint, it does not mean you can build a house." The guidance of an experienced Lean coach will substantially improve the likelihood of building a map that clearly illustrates wastes and of envisioning a future state where the appropriate Lean processes have been utilized to reduce or eliminate those wastes.

Summary

The primary purpose of a VSM is to depict flow through the process in as clear and lucid a manner as possible, so that it becomes apparent how we are adding value to the product to satisfy customer needs, as well as what we are doing that consumes time or resources and doesn't add to customer value. Anything that can be done to enhance the clarity and visibility of the map so that the messages are more obvious should be done. We have touched on a number of those: always showing flow in a straight line from left to right, basing the map on logical flow rather than trying to be geographically accurate, and not cluttering the map with unnecessary detail.

You will see concrete examples of those principles in action as we draw a real process VSM, so it's time to introduce the process that will provide our focusing problem.

Chapter 5

Our Focusing Problem— A Synthetic Fiber Process

Process Overview

To begin to draw a real VSM, we will examine a synthetic fiber making process in Riverside, North Carolina owned by the Carolina Fiber Company. This process makes fibers in two forms—very short filaments (1/2 to 2 in., sometimes called staple) sold to customers who manufacture high-end carpets, military garments, and high-end sporting apparel, and as thick ropes (thousands of feet long, often referred to as tow) sold to customers who do their own processing before blending and weaving. The short filament product, the staple, is sold in large bales, similar to cotton bales, and comprises about 80% of the total sales volume. The long rope product, the tow, is sold in large cardboard boxes called Gaylords, and represents about 20% of sales.

Annual customer demand for the baled staple is 17,600,000 lb. Tow demand is 5,000,000 lb per year.

The process runs 24 hours per day, 7 days per week, including holidays.

There are four major steps in the process.

1. Polymerization (*a chemical process*)
2. Fiber extrusion, called spinning (*a mechanical process*)
3. Drawing (stretching) and annealing (heat setting) (*mechanical processes*)
4. Fiber cutting and baling (*a mechanical process*)

Product demand is slightly variable, with no recognizable seasonal trends. Demand for fibers used in military uniforms goes up and down with the level of international conflict.

Raw Materials

There are two primary purchased raw materials, adipic acid and hexamethylene diamine, delivered in railcars of 100,000 lb each. De-ionized water is produced on-plant, and is always available in any reasonable quantity.

Raw materials are stored on-plant in the railcars. There are typically 3 to 4 railcars of each material on-site at any time.

Supply lead time is 7 days, from order placement or material release to receipt of railcars. Each supplier takes about 1 day to process the order and fill the railcar from existing inventory, and transportation time is 6 days. The supplier is very reliable. Standard deviation of lead time is less than 1 day.

At the start of each month, each supplier is given a schedule of dates to release railcar orders, based on the month's production schedule.

Step 1: Polymerization

Three raw materials (adipic acid, diamine, and de-ionized water) are mixed and heated under pressure to form a very viscous plastic, called a polymer. This polymerization occurs in large jacketed reactors. There are three independent reaction systems. A typical batch is 760 gallons; density is 7 lb per gallon. Charging raw materials into the reactor takes 15 min, polymerization takes 90 min, discharging and extrusion takes 30 min.

Twelve different materials are produced with different viscosities. Each of the 12 materials is produced once every 8 days, in one of the three reactors.

There is no cleaning between batches of the same material. A 20-min flush is required between batches of different material.

Each reactor is taken offline once per year for a 48-hour disassembly and wall cleaning.

The polymer is extruded as a group of ribbons, which are then cut into small pellets, cooled, and stored in large silos. These pellets are often referred to as Flake. There are 12 silos, one for each type of polymer, with a 200,000-lb capacity each. The contents of each silo go up and down during the production cycle, but the average total across all 12 silos is relatively constant at 600,000 lb.

The process is very stable and predictable, with a yield of 97%. The reliability of the reaction system, including the reactor and all piping, pumps, and sensors and controls, is 94%.

Two control room operators run the three reactors each shift. No lunch relief is needed.

Step 2: Fiber Spinning

The plastic pellets are pneumatically conveyed from one of the silos to a spinning machine, which melts the pellets and extrudes them as continuous strands (filaments) of fiber. As the filaments are being extruded, cold air is blown on them to quench and solidify them. An extrusion head ("spinnerette") has 120 orifices, so it extrudes 120 filaments in parallel, which are gathered together in a bundle. A spinning machine has 96 to 144 spinnerettes, each of which spins a bundle of 120 filaments. The output from all of these spinnerettes is gathered together to form a very thick rope-like bundle of filaments. Thus, the rope consists of about 15,000 (11,520 to 17,280) of these very fine filaments. Fiber is extruded at a velocity of 1000 yards/min.

The fiber rope is continuously guided into a large (5' × 5' × 5') stainless steel tub. Each tub can hold approximately 1000 lb of fiber. As a tub becomes full, the fiber is automatically cut and guided into the next tub in a line. The full tubs are taken to a storage area, to be staged for the next process step, Draw–Anneal. The storage area holds 800 tubs. It is typically 80% full.

There are six identical spinning machines in parallel. Each machine is capable of producing 660 lb of fiber per hour if running perfectly.

Sixty different products are produced, with different fiber thicknesses, dye ability, and modulus; each product is made once every 7 days. It takes 2 hours to reconfigure a spinning machine to produce the next product. While this is happening, the molten polymer is being spun to waste. However, the rate can be cut slightly so that the material lost on a changeover is only approximately 1000 lb.

Spinning yield is 88%, reflecting primarily the material wasted on product changes. Electrical–mechanical reliability is 93%

There is a single operator for each spinning machine; operators help each other out with tasks that require two people.

Step 3: Draw–Anneal

Tubs of spun fiber are moved from the tub storage area to one of five draw machines. The fiber rope is pulled from the tub and fed through several sets of nip rolls running at successively faster speeds, thus stretching the fiber bundle. This stretching ("drawing") causes the long-chain molecules to orient and align, and increases tenacity and other properties.

After drawing, the rope continues through a steam chamber, where the properties are heat set ("annealed") to stabilize them. (Four of the five machines use steam to anneal the fiber; one uses hot rolls, described next.)

The rope-like bundle is then fed into large containers called cutter boxes, or into large cardboard boxes ("Gaylords"). Each cutter box or Gaylord holds approximately 800 lb of fiber. Products sold as rope (10 SKUs) are fed into the Gaylords, while products to be cut and baled (98 SKUs) are put into the cutter boxes. The cutter boxes are moved to a lag storage area, waiting cutting/baling. The area typically has 375 cutter boxes stored. The Gaylords are moved into finished product storage, which typically has approximately 172 Gaylords stored as finished product.

Each steam anneal machine can process 1100 lb per hour on a perfect day. Yield is 90% and reliability is 92%. Product changes can be done almost instantaneously.

The one machine that uses heated rolls rather than steam to accomplish the annealing is an older machine, with poorer reliability (78%), lower yield (85%), and lower maximum throughput (770 lb per hour). The hot roll temperature typically must be changed on product changeovers, taking about 30 minutes to stabilize at the new temperature.

All draw–anneal machines run a one-week campaign cycle, or EPEI.

Eight SKUs, which comprise about 10% of the total annealed output volume, must be run on the hot roll anneal machine. These products would likely run acceptably well on the steam anneal machines, but significant process development and product prove-out and qualification would have to be done. All products requiring hot roll annealing are sold cut and baled, none as rope.

Because of lower demand for hot roll products, that machine is run 24 hours per day, but only 5 days per week.

Each draw–anneal system has a single operator assigned per shift. There are tasks requiring two operators, so they help each other out on those tasks.

Step 4: Cut—Bale

Cutter boxes of drawn fiber are transported from the lag storage area to the input end of one of the four cutter/balers. Eight cutter boxes are positioned to feed the cutter/baler in parallel through a set of yarn guides called a creel. Fiber is simultaneously pulled from all the boxes in the creel and fed into the cutter/baler.

The rope bundle is cut into very short fibers (1/2 in. to 2 in.) by a rotary cutter mechanism with razor blades positioned every half inch (or more) around the circumference. The cut fibers fall into a chamber. When the chamber is full, the walls move inward to compress the mass of fiber into a bricklike form. The brick is ejected to a wrapping section, where the material is wrapped with Typar under pressure. Plastic strapping is applied to the wrapped bale to help contain it. The finished bales are ejected onto a conveyor, where they are conveyed to a finished product storage area. The storage typically contains 900 bales.

When running ideally, a baler can process 1200 lb per hour. Each bale weighs 770 lb.

Due to the complex mechanical structure and linkages, the high pressures used, and the tendency for fiber dust and fines to get into bearings and linkages, reliability is poor. It varies slightly from baler to baler, and averages 76%. Baler yield is 98%.

There are 180 baled products, but some are very low volume, and very infrequently sold products. Only about 120 are made on a regular basis. The balers are campaigned on a 7-day basis.

Product changes require about 2 hours. All fiber from the previous run must be cleaned out so that there is no product contamination. Getting 1-in. fiber into a 2-in. bale is not much of an issue, but the reverse is. Similarly, getting thinner filaments into a bale of thicker fibers is not much of a problem, but the reverse is. Nonetheless, contamination by mixing filament thickness is not as serious as mixing cut lengths.

Each baler is assigned one operator per shift, to stock the creels and to monitor baler performance.

Finished Product Storage and Shipping

Bales and Gaylords are stored waiting customer orders. Each is sold in truckload quantities. When an order is received, the bales or Gaylords are loaded onto the trailer by fork trucks or clamp trucks. A full trailer holds 40 bales or 40 Gaylords. It takes anywhere from 90 minutes to 4 hours to load one trailer, including preparation and checking of the shipping manifest, depending on the number of material handlers available.

The customer lead-time requirement varies with the customer and with product type, but in most cases is 4 days. Average ship time from the Riverside plant to customers is 2 days. Therefore, the truck must leave the site within 48 hours of order receipt, so the goal is to have enough material in finished product inventory to satisfy all orders. That requires our production to follow a make-to-stock strategy.

There are four material handlers per shift to move tubs, cutter boxes, bales, and Gaylords, and to load shipping trucks.

Order Processing and Production Scheduling

Customers call Customer Service Representatives (CSRs) to place orders. The CSRs check the current inventory to see if the material is available. If it is, the CSR releases the material for shipment and confirms the shipment with the customer.

Based on input from customers and historical trends, CSRs prepare sales forecasts for the next quarter. All CSR forecasts go into a Demand Management process for the entire business. Based on this input, other inputs, and historical trends, a forecast for the next month is prepared. This forecast is input into the monthly Sales and Operations Planning (S&OP) process.

The plant Production Supervisor prepares a Capacity Forecast for the coming month, incorporating any planned shutdowns, clean-ups, and test production runs. The S&OP process reconciles the demand forecast with available capacity and generates a Production Schedule for the coming month.

The Production Schedule is used to calculate raw material needs for the month. This triggers orders to the suppliers to cover the coming month.

Based on the Production Schedule, the Production Supervisor issues a daily production plan for the next seven days each week. The daily production plan may be updated each day as needed to reflect the previous day's performance.

The Synthetic Fiber Manufacturing VSM

The Riverside plant is facing intense and ever-increasing competitive pressure, very typical of the domestic fiber industry since the advent of Asian competition, so it must make its operations as effective as possible and eliminate all waste. This is the same pressure that drove Toyota to begin to develop Lean 60 years ago, but in that case the pressure was felt in Asia coming from the West, where now it is coming from the opposite direction.

In order to deal with these pressures, Riverside has decided to undertake a full Lean transformation, and has wisely chosen Value Stream Mapping as one of the first steps.

The full VSM of its process is shown in Figure 5.1. In the next few chapters, we'll see how it created this map, step by step from the process described previously.

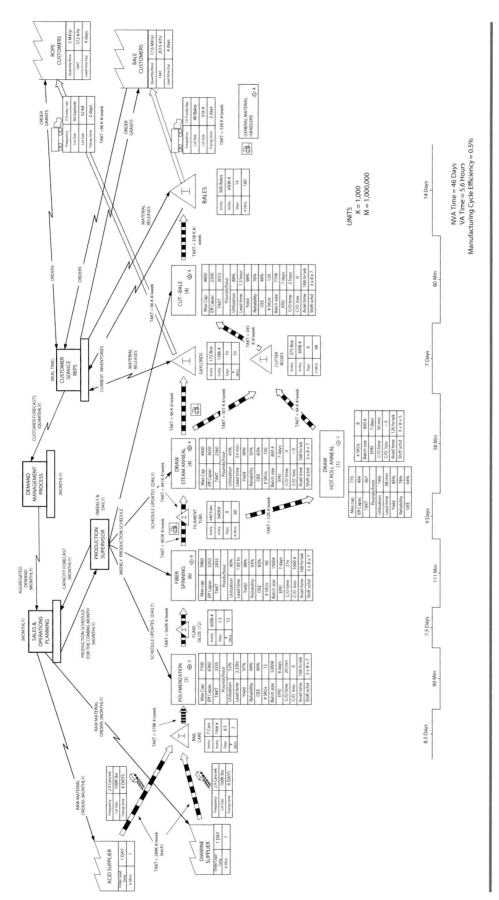

Figure 5.1 A VSM of the fiber manufacturing process.

Chapter 6

Developing the Material Flow

The material flow is actually one of the easier portions of the VSM to develop. If you have an accurate understanding of the process (as we provided in the previous chapter), capturing the material flow on paper or electronically is reasonably straightforward.

As we noted in Chapter 3, you can begin at the raw material input and work forward through the process flow, or start with customer orders and work backward from the shipping dock. The latter is in keeping with traditional Lean guidance as it begins with a focus on the customer. However, the former seems more intuitive for mapping teams, particularly when mapping process lines, so that's the approach we'll follow here. Later, when we get to the Takt calculations, we'll start with the customer and work back.

Figure 6.1 shows the completed material flow portion of the VSM.

We will start with raw material supply. There is a single supplier for each of the two main ingredients in our process, so they are shown as two supplier boxes. We could have shown one supplier box, with a quantity of two, and that would have been just as appropriate. The two materials are supplied in approximately equal quantities, with equal delivery times and lead times, so there is nothing requiring them to be differentiated on the map. We thought it would add useful information to the map by showing two boxes. This is one of those situations where there is more than one correct way to represent something on the map, and so the choice is up to the judgment of the mapping team.

The raw material inventory in railcars on the plant is shown by the typical inventory icon, a triangle with the capital letter "I" inside.

There are three polymerization reactors, and there is nothing to suggest that they are different in any meaningful way, so one process box is shown with a quantity of 3.

Another inventory triangle is shown to represent the flake storage in silos, with a notation that there are 12 silos. The quantity typically stored will be covered later when the data boxes are added.

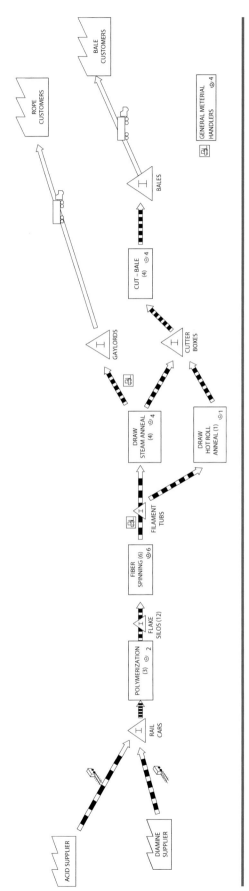

Figure 6.1 The material flow portion of the VSM.

Fiber spinning is next, and with nothing to differentiate the six spinning machines, they are shown as a single process box.

After fiber storage in filament tubs, we come to the draw–anneal step, where we encounter the first process step with significantly different kinds of equipment in parallel. As discussed in Chapter 5, the hot roll annealer is sufficiently different from the steam annealers, both in operating characteristics and in product requirements, to require showing it as a separate process in parallel.

Note that draw–steam anneal is actually two process steps, two separate physical transformations performed in sequence. However, because there is continuous flow from Draw to Anneal, and parameters like yield are only known for the combined steps, we show them as one step. Similarly, draw–hot roll anneal is actually two sequential processes shown as one. If a material leaving draw–anneal doesn't have the specified properties, it is not always clear whether it is because of improper drawing conditions or incomplete heat setting. Thus, from a flow viewpoint, there is no benefit in differentiating them. For the purposes of understanding waste and its causes, there may be. However, we can explore that with a more detailed process map of those specific steps.

We have two types of inventory to show coming out of the draw–anneal step. The product to be sold as tow, i.e., rope, is stored in Gaylords, which is the container medium used to ship it to customers. We know from the process description that the product types sold as tow are all annealed on the steam anneal machines. Other products from steam anneal go into cutter boxes, as do all of the product types from the hot roll annealer.

Cut–bale is next. The four machines have slight differences, but none significant enough to cause them to be shown separately. Therefore, they are shown as one process box with an indication that there are four machines in parallel. The bale finished product inventory is shown with the conventional triangle with an "I" inside.

The bale customers and rope customers could be shown by one customer box, but because the product families are significantly different, we chose to show them as two customer boxes. If a plant produces one general type of product with a number of variations, it is generally shown with a single customer box. However, if the plant produces product families that are considerably different, they can be shown with different customer boxes. Anytime useful information, information that creates a clearer picture of flow, can be shown on the map without adding clutter it should be done.

All of the arrows connecting process steps and inventories are shown as crosshatched arrows to signify that the system is based on "push" scheduling concepts. The term "push" is used to characterize a material replenishment strategy where material is pushed through the process based on forecasts without regard to current inventories and orders. We learned from Chapter 5 that all fiber production is based on the monthly forecast, without taking current inventory into account, so this is clearly a push system. This is in contrast with "pull" where material is replenished only to replace material that customers (or the next step in the process) have pulled from inventory. Replenishment in a pull system

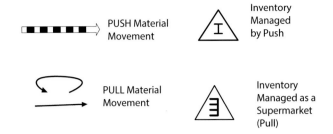

Figure 6.2 Push and pull icons.

is based on real-time conditions on the plant floor, and production is allowed only when there is a need to refill inventory or satisfy current customer orders. Pull is considered a superior replenishment strategy in that it avoids overproduction, i.e., producing more than is currently needed, and manages inventory to reasonable, predictable levels. The VSM convention is to show push flow by a crosshatched arrow, and pull by a narrow line arrow as shown in Figure 6.2 (more detail on pull concepts and benefits can be found in Appendix D).

For the same reason, the inventories are all shown by a triangle with the capital letter "I," which represents an inventory filled by push methods. When an inventory is refilled according to pull principles, the triangle has a symbol resembling a supermarket shelf. The origin of that is that manufacturing pull concepts were inspired by the methods used to refill supermarket shelves in the 1950s.

With the material flow portion of the map now complete, it's time to turn our attention to the data boxes.

Chapter 7

Calculating Data Box Parameters

Process Step Data Boxes

Most data box parameters, such as Overall Equipment Effectiveness (OEE), can be calculated simply from what you know about that particular process step. However, parameters based on Takt, such as utilization and days of supply, can't be calculated until Takt has been determined for each step. As was mentioned in Chapter 2, Takt can be different from step to step to make up for yield losses and because available time may be different for some steps. Therefore, Takt must be calculated on an integrated basis for the entire process, and we must start from the customer demand and build it backward into the process. This is the one aspect of a VSM where we must start with customer requirements and work back through the process flow.

In this chapter, we will calculate all data box parameters that are independent of Takt; Takt and the parameters that depend on Takt values will be calculated in the next two chapters.

Cut–Bale (Figure 7.1) will be used as an example to illustrate process data box calculations.

The process description told us that each baler can process 1200 lb per hour (pph), so the Maximum Capacity is 4800 pph for the group of four balers.

Effective Capacity depends on OEE, which will soon be calculated to be 48%, so:

$$Effective\ Capacity = Maximum\ Capacity \times OEE = 4800 \times 48\% = 2300\,pph$$

Takt, and Utilization which depends on Takt, will be calculated in subsequent chapters.

CUT – BALE	
(4) ↻ 4	

Max Cap	4800
Eff Capac	2300
TAKT	
Pounds/hour	
Utilization	
Lead time	1.3 hour
Yield	98%
Reliability	76%
OEE	48%
# SKUs	120
Batch size	770#
EPEI	7 days
C/O time	2 hour
C/O loss	0
Avail time	168 hr/wk
Shift schd	3 × 8 × 7

Figure 7.1   Cut–Bale process step data box.

A bale weighs 770 lb, and each baler has an Effective Capacity of 575 pph, so the time to produce one bale is:

$$Lead\ Time = \frac{Batch\ Size}{Effective\ Capacity} = \frac{770\ lbs}{575\ PPH} = 1.3\ hours$$

So *on average*, each baler ejects a bale every 1.3 hours. (If running perfectly, each baler could produce a bale every 39 minutes between changeovers.)

Overall Equipment Effectiveness (OEE)

Next is OEE, one of the most widespread measures used to gauge how well your equipment is performing. One reason for its popularity is that it captures all of the factors that detract from optimum equipment performance in a single metric. OEE is the product of three factors:

- Availability
- Performance
- Quality

Availability captures all downtime losses, including breakdown maintenance, minor stops, preventative maintenance, and time spent in changeovers. Availability is calculated as actual operating time divided by planned production time.

$$Availablilty = \frac{Actual\ Operating\ Time}{Planned\ Operating\ Time}$$

Performance captures the loss in productivity if equipment must be run at less than the design throughput rate because of some equipment defect. For example, chemical batches can take longer to heat up or react if residue has built up on vessel walls, thus impeding heat transfer. Rotating machinery, paper winding equipment, or plastic film processing equipment may have to be run at slower speeds if bearings are worn. Performance is calculated as actual throughput divided by rated throughput.

$$Performance = \frac{Actual\ Throughput}{Rated\ Throughput}$$

Quality captures the loss in equipment productivity when out-of-specification product is being made, including material that must be reworked to be acceptable. Yield losses during restart coming back from a product changeover are not included in Quality because they have been accounted for in the total changeover time component of Availability.

$$Quality = \frac{Quantity\ of\ First\ Grade\ Material}{Total\ Quantity\ Produced}$$

OEE is then calculated as:

$$OEE = Availability \times Performance \times Quality$$

Calculating Availability

Availability is a function of reliability (which is based on the amount of time a machine is broken down or undergoing maintenance) and changeovers. Applying this to the balers, we know that the reliability is 76%. The reliability number means that a baler is down for maintenance 24% of the time we are trying to operate it, and here's where the OEE calculation can get tricky; the reliability factor applies only when we are trying to run, so the 24% downtime is not figured on the portion of time spent in changeovers.

We know that the balers must collectively bale 120 products, so each baler must process about 30 products. Since a changeover takes 2 hours, a baler spends 60 hours per week involved in changeovers. Therefore, a baler is actually operating only 108 hours (168 – 60) in an average week, and the 24% maintenance downtime should be calculated from that time. Thus, the reliability downtime is 24% of 108 hours, or 26 hours. The total time to be detracted for Availability is 60 hours + 26 hours, and the actual operating time is 82 hours (168 hours – 86 hours).

$$Availablilty = \frac{Actual\ Operating\ Time}{Planned\ Operating\ Time} = \frac{82\ hours}{168\ hours} = 49\%$$

(This assumes that reliability is measured as the downtime divided by the time we are trying to run. If the reliability factor is calculated based on percentage of available time, then that portion of time lost is 24% of 168 hours, and the availability factor in OEE is 40%.)

Calculating Performance

There are no equipment problems forcing any baler to run at less than the design rate, so Performance = 100%.

Calculating Quality

The Yield is 98%. These are all losses of the very fine fibers that adhere to the walls of the machinery as the material flows through. These are cleaned out during a changeover; there are no additional losses during changeovers.

Calculating OEE

The resulting OEE factor is:

$$OEE = Availability \times Performance \times Quality$$

$$OEE = 49\% \times 100\% \times 98\% = 48\%$$

Uptime is another metric often used to gauge equipment performance. It measures the same losses that OEE does, but performs the calculations in a different way. Even so, it gives the same result as the OEE calculation.

Remaining Factors

The number of baled SKUs run on a routine basis is 120, and the Batch Size is one bale, 770 lb.

Every one of the 120 products routinely produced is made every week, for an EPEI (Every Part Every Interval) or overall campaign cycle of 7 days. As noted previously, the changeover time averages 2 hours, and there are no materials lost on changeover.

The Shift Schedule is three 8-hour shifts per day, 7 days per week, for an Available time of 168 hours.

FIBER SPINNING (6) ↻ 6	
Max cap	3960
Eff Capac	3250
TAKT	
Ponds/hour	
Utilization	
Lead time	1.85 hr
Yield	88%
Reliability	93%
OEE	82%
# SKUs	60
Batch size	1000#
EPEI	7 days
C/O time	2 hr
C/O loss	1000 #
Avail time	168 hr/wk
Shift schd	3 × 8 × 7

Figure 7.2 Fiber spinning process step data box.

Another Example of OEE

The OEE calculation for Fiber Spinning (Figure 7.2) is worth mentioning because the changeover time and the yield losses overlap. Essentially all of the 12% yield loss is the material wasted during the changeovers, and comes from the need to keep material flowing during the changeover. If polymer flow were to be stopped, it would freeze in the piping and pumps, and require everything to be disassembled and blasted or burned out. It is typical of continuous polymer processes that material flow cannot be interrupted. In this case, the rate can be throttled back during a changeover, from the normal flow of 660 pph to 500 pph, so the total loss on any 2-hour changeover is 1000 lb instead of 1320 lb.

There are 60 products run on the six spinning machines, about 10 products each. Thus, there are ten 2-hour changeovers on each machine each week. Therefore, 20 hours are lost to changeovers each 168 hours, and the time actually spent trying to spin fiber is 148 hours. Applying the 93% reliability factor to that indicates that a spinning machine is typically down for maintenance 10 hours (7% × 148 hours). The actual operating time is 138 hours (168 − 20 − 10), so Availability is

$$Availablilty = \frac{Actual\ Operating\ Time}{Planned\ Operating\ Time} = \frac{138\ hours}{168\ hours} = 82\%$$

There are no equipment caused rate reductions, so Performance is 100%. Quality can be considered to be 100% for this calculation because the only Quality losses are on restart from a changeover, which is accounted for in changeover time.

$$OEE = Availability \times Performance \times Quality$$

$$OEE = 82\% \times 100\% \times 100\% = 82\%$$

To emphasize an important point, a spinning machine is capable of making good product 82% of the time. It is true that there are significant quality losses (which will be addressed later), but they occur during the time lost in change-overs, so their effect on throughput is accounted for in that factor.

Supplier Data Boxes

Supplier data boxes are typically very simple. In the case of the adipic acid supplier (Figure 7.3) the Order Lead Time, the time from when the supplier receives the order or material release to the time the material leaves his plant, is 1 day. The time it takes the railcar to reach our plant will be accounted for in the transportation data box.

We purchase only one grade of acid, so the SKU count is 1.

Customer Data Boxes

Customer data boxes are also straightforward. For bale customers (Figure 7.4) we know that the total annual sales volume for the baled products is 17,600,000 lb per year. From the 168 hours per week of available time, and 52 weeks per year operation, the Takt for this product family is 2015 pph. This also could have been listed as 2.6 bales per hour.

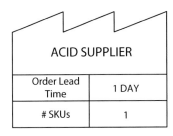

Figure 7.3   Supplier data box.

Figure 7.4   Customer data box.

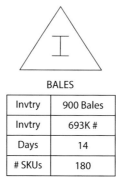

BALES

Invtry	900 Bales
Invtry	693K #
Days	14
# SKUs	180

Figure 7.5 Inventory data box.

If the customer Lead Time requirements are known, they should be shown in the data box. In this case, the Lead Time requirements vary from customer to customer but are 4 days for most customers, so that is what is listed.

Inventory Data Boxes

Inventory data boxes are very important because inventory is typically one of the most significant wastes in any manufacturing operation, and much more so in a process plant. Therefore, it is critical that this waste be presented in a highly visible way. We know that bale inventory (Figure 7.5) averages about 900 bales. At a weight of 770 lb per bale, that comes to 693,000 lb stored.

The number of days that this represents can be calculated using Little's Law:

$$Lead\ Time = \frac{Total\ inventory}{Throughput\ rate}$$

$$Lead\ Time = \frac{693,000\ lbs}{2015\ pph} = 344\ hours = 14\ days$$

The number of days of supply is very useful because it puts the amount stored in a meaningful context; it is more revealing to know that bale inventory is 14 days than it is to know it is 693,000 lb or that it is 26 turns.

The number of bale types stored as finished product is 180, the 120 with reasonable sales volumes and the 60 slow movers. (The Cut–Bale data box shows 120 SKUs because that is the number produced on a routine basis, but the inventory box shows 180 because most of the slow moving and obsolete products have some inventory.)

Transportation Data Boxes

The transport of adipic acid railcars from the acid supplier to our plant provides a good example of a transportation data box (Figure 7.6). We know that a railcar

Frequency	
Lot Size	100K lbs
Transp time	6 DAYS

Figure 7.6 Transportation data box.

holds 100,000 lb. The total supply lead time is 7 days, and the transportation portion of that is 6 days.

We can't yet calculate the frequency at which the railcars are shipped because we don't yet know the Takt, the required flow rate, at this point in the mapping process, so that will wait for later.

Summary

We've done most of the work to populate the data boxes, but have to tackle the most important component—Takt. It is the parameter that ties everything together, and turns a collection of independent, discrete steps into an integrated process. It is the parameter that describes flow requirements, and from that, we can determine how well we can achieve flow.

Chapter 8

Material Flow Rates and Takt

Calculating Takt

Most of the data in the data boxes can be calculated on an individual, step-by-step basis, as we saw in the last chapter. The one parameter that can't be computed on a stand-alone basis, the thing that ties it all together, is Takt, which must be calculated and analyzed on an integrated basis. Takt defines the flow requirements, and therefore the actual flow, in a well-synchronized process. A key benefit of VSMs is that they tie all steps together as an integrated whole to determine flow and capacity requirements, and clarify how the performance of one step influences the requirements of other steps.

The benefit of calculating Takt as a rate-based parameter rather than a time-based parameter for most process operations was discussed in Chapter 3, and that's the way we'll do it here.

Takt values can be calculated in a time frame of hours, days, weeks, or years. When looking at total demand from all customers for all baled products, it is practical to record it in terms of millions of pounds per year. When looking at the Takt that any specific piece of equipment such as a baler must process, it may be more appropriate to talk in pounds per hour because that's the way we typically think of baler capacity. When calculating the Takt requirements for the flows through all of the process steps and process equipment, looking at weekly time increments may make the most sense. In the synthetic fiber process, for example, the two types of draw anneal equipment have different operating schedules and different numbers of hours per week. Looking at Takt on a pounds per week basis puts everything on a common basis, which makes it easier to see how the Takt numbers relate to each other and add up. (This is a concern only when different pieces of equipment have different available times.) Therefore, for the fiber process, we will use pounds per year for the aggregate customer demand, pounds per week to trace the flow requirements through the system, and pounds per hour for each specific piece of equipment or equipment group.

Instead of pounds, we could use bales and Gaylords as our unit of production quantity for the final process steps. However, a pound is the unit typically used for customer orders; even though customers receive Gaylords or bales, what they order is in terms of pounds. Moreover, a pound gives us a unit that is consistent throughout the process, so that is what we'll use.

To do all the Takt calculations, we'll start with the customer demand and trace flow backward through each step in the process.

Bales

The aggregate sales of all baled products to all bale customers are 17,600,000 lb/year. That means that the balers must collectively produce 338,000 lb/week, and with 168 hours of available time each week the four machines must cut and bale 2015 pph.

The 98% baler yield shown in the data box in Figure 8.1 means that 2% of the baler input goes to waste, so the Takt of drawn fiber coming from the cutter box inventory is 345,000 lb/week.

Rope Takt in Gaylords

The aggregate sales of all rope products to all customers are 5,000,000 lb/year. As shown in Figure 8.2, this gives a Takt of 96,000 lb/week and 572 pph for the rope finished products packaged in Gaylords.

Annealed Product Takt

Figure 8.3 depicts the flows to be taken into account to calculate Takt through the draw–anneal machines. The five machines combined must produce 96,000 lb/week in Gaylords and 345,000 lb/week in cutter boxes, for a total of 441,000 lb/week. We know that 10% of that must be annealed on the hot roll draw–anneal machine, or approximately 44,000 lb/week. That leaves 301,000 lb/week to be annealed on the steam anneal machines to go into cutter boxes and 96,000 lb/week to go into Gaylords. Thus, these machines must produce 397,000 lb/week and with an available time of 168 h/week, the steam anneal Takt is 2360 pph.

The hot roll annealer runs only 5 days per week, so its available time is 120 h/week and its Takt is 367 pph.

If we had used pounds per hour (pph) based on a 168-hour week rather than pounds per week as the common unit to trace Takt back through the process, the hot roll anneal machine would have an apparent Takt of 262 pph. We would have had to adjust that by a factor of 168/120 to account for the machine having only 120 hours of available time per week to get to the true Takt of 367 pph.

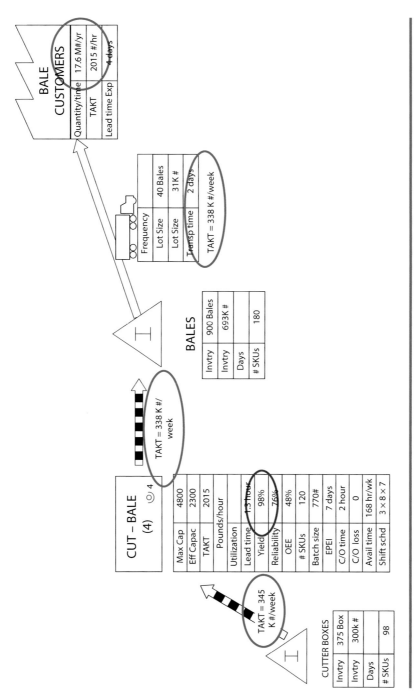

Figure 8.1 Bale and cutter box Takt.

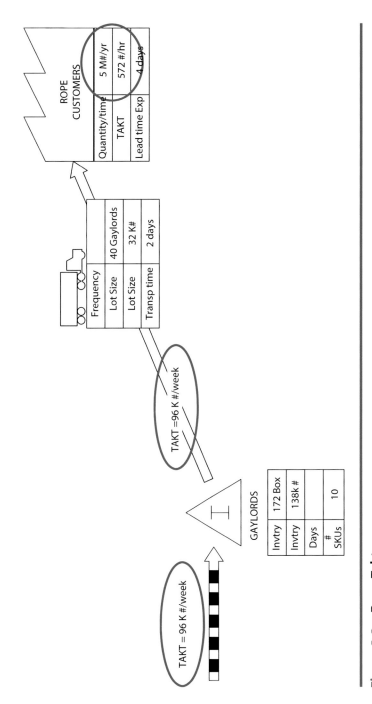

Figure 8.2   Rope Takt.

That's why it is more straightforward to use pounds per week as the Takt parameter for any process that has equipment with different available times.

Filament Takt

Figure 8.3 also illustrates the factors involved in determining filament Takt. The four steam annealers have a combined Takt of 397,000 lb/week, and so must be fed 441,000 lb/week to make up for the 10% yield loss. The hot roll anneal machine must output 44,000 lb/week, and so must consume 52,000 lb/week to account for the 15% yield loss on that machine. Thus, the total Takt of filament product coming from the spinning machines is 493,000 lb/week (441,000 + 52,000).

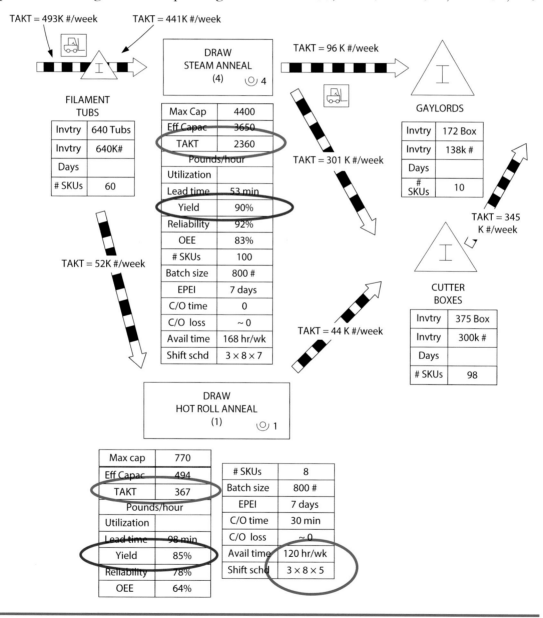

Figure 8.3   Annealed product Takt.

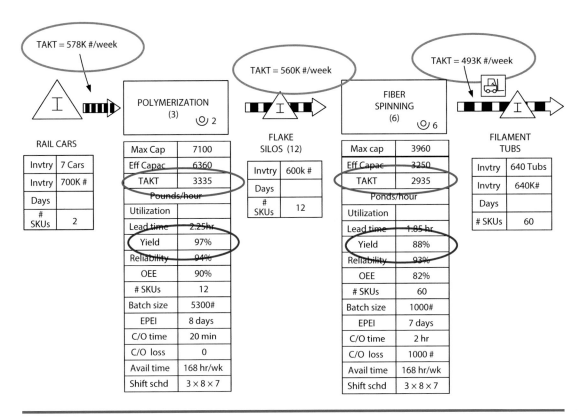

Figure 8.4   Flake, spinning, and polymer Takt.

Flake, Spinning, and Polymer Takt

The 493,000 lb/week demand from the spinning machines translates to a spinning Takt of 2935 pph, based on available time of 168 h/week. With a spinning yield of 88%, the flake Takt becomes 560,000 lb/week. Thus, the three polymerization reactors must produce to that Takt, or 3335 pph, as illustrated in Figure 8.4.

Raw Material Takt

There is a 3% yield loss in polymerization, so the combined Takt of all raw materials is 578,000 lb/week. Adipic acid and diamine constitute almost all of the volume required for polymerization, and are consumed in approximately equal quantities, so the Takt of each is 289,000 lb/week as illustrated in Figure 8.5.

Summary

Now that the Takt values throughout the entire fiber manufacturing process are known, we can complete all the data boxes by calculating the parameters that depend on Takt: utilizations for all process steps, days of supply for each inventory, and frequency for each transportation step, and that will complete the material flow portion of the fiber VSM.

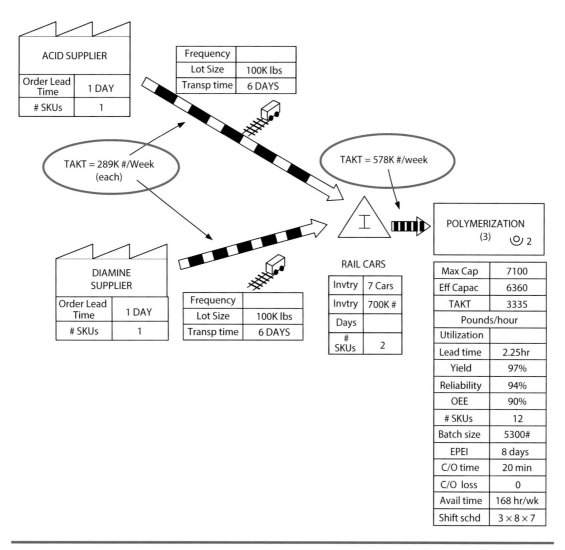

Figure 8.5   Raw material Takt.

Chapter 9

Completing the Data Boxes

*Utilization, Delivery Frequency,
and Days of Supply*

In the last chapter, we calculated Takt throughout the process. Now all the parameters that depend on Takt can be calculated—specifically, utilization in process steps, delivery frequencies in transportation, and days of supply for inventories.

Utilization

Utilization is an important VSM parameter and can highlight potential flow barriers as well as the waste of asset capacity. A process step with high utilization can be a potential bottleneck, and the higher the utilization the more likely that it will be a bottleneck. A bottleneck is any step in a process with an effective capacity less than Takt, that is, a process step that does not have the capacity to meet customer demand. A step with utilization lower than 100% is theoretically not a bottleneck, but can be during periods when reliability or yield are at the low end of their variability profile. A machine with 90% reliability could have a perfect day followed by a day where it is down 20% of the time. So even though the utilization, which is based on average OEE values, may be less than 100%, it may exceed 100% for periods of time. Thus, some degree of headroom in the utilization is desirable. Likewise, demand can vary from day to day, and Takt values are based on averages. But this is less of a concern because the production scheduling methods we will adopt in the future state offer production leveling benefits.

Production managers often ask what level of utilization is reasonable, what utilization value is "safe." Utilizations of 85% to 90% are reasonably comfortable because reliability problems and yield losses have already been accounted for in Effective Capacity. However, 95% to 100% is not practical because those OEE factors have variability.

While high utilizations can be a sign of potential problems, low utilizations can be a sign of opportunities; low utilizations on parallel equipment point to equipment that could possibly be mothballed to save operating and maintenance cost. But this should be analyzed carefully. You don't want to sacrifice flexibility and Group Technology opportunities without careful evaluation. Group Technology is the name sometimes given to the practice of dedicating product families to specific pieces of parallel equipment to reduce changeover time and cost. Group Technology and Cellular Flow are explained in Appendix C.

While Utilization is a very useful indicator of potential problems and potential improvements, it is not a KPI (Key Performance Indicator) to be maximized! Some production managers have a metric called utilization that is based on actual production rather than Takt, and try to maximize this so-called utilization for the sake of being able to report a higher number. This may be seen as improvement, but leads to overproduction and more inventory than you need. A much better metric is "Performance to Plan," which measures how well you are producing whatever the plan calls for, ideally Takt, and penalizes overproduction as well as underproduction.

Turning to Figure 9.1, we can calculate the utilizations for all the process data boxes in the fiber process. Cut–Bale has four machines with a combined effective capacity of 2300 pph and a Takt of 2015 pph, so its utilization is 88%. This means that each baler must run about 88% of the time left after changeover times, downtimes, and the effective time lost in making off-specification product have been deducted. This is a reasonable value, not overly high or low. If we want to lower it to give more operating headroom, then the poor reliability would be the likely place to start.

The four draw–steam anneal machines have a utilization of 65%, based on a Takt of 2360 pph and an effective capacity of 3650 pph. This relatively low utilization may provide the opportunity to move the hot roll annealed products to the steam anneal machines so that the hot roll machine can be eliminated.

The hot roll machine, with a Takt of 367 pph and an effective capacity of 494 pph, has a utilization of 74%. If we decide to keep it in operation, this low utilization number indicates that we could consider operating it even fewer shifts each week.

Fiber spinning has a utilization of 90%, which is high, but not high enough to be a significant concern. The bigger concern in spinning is the 2-hour changeover time and resulting loss of large amounts of valuable materials. This should be addressed, and if improved, will lower the utilization as a collateral benefit.

This brings us to the three polymerization reactors. The utilization here is very low, 52%, and strongly suggests that two reactors can handle the load. The energy savings resulting from taking a large, heated chemical vessel offline are very substantial, and certainly not the only savings possible, so this should be thoroughly evaluated. This will be done when we get to VSM analysis in Chapter 12.

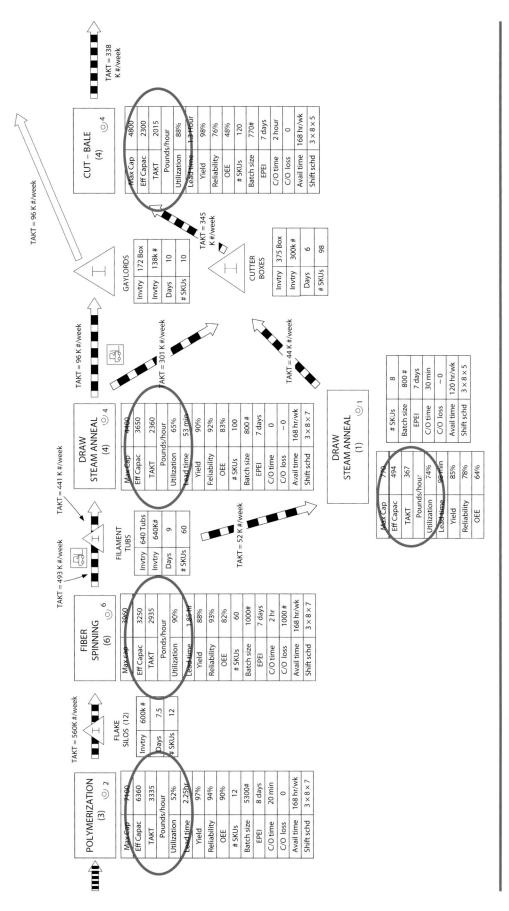

Figure 9.1 Process step utilizations.

Transportation Frequency

Transportation frequency is another parameter based on Takt that can now be calculated. Delivery frequency can be a good indication of opportunity: if frequency is low, it may be possible that smaller lot sizes could be delivered. Smaller, more frequent lot sizes will lower inventory at raw material input: smaller delivery quantities will reduce raw material cycle stock, and deliveries that are more frequent will reduce the time at risk and therefore safety stock. More frequent deliveries of smaller quantities to your customers will offer them similar inventory benefits and may improve customer relationships. It is also possible that smaller lots will smooth out variability in customer demand.

Figure 9.2 shows the customer delivery portion of the VSM material flow. Looking at bale shipments, the typical shipment is one truckload, weighing 31,000 lb. With a Takt of 338,000 lb/week, we ship about 11 trucks each week, or 1.6 trucks per day. Similarly, deliveries to rope customers are 32,000 lb per truckload, so a Takt of 96,000 lb/week requires three truck deliveries per week. Because customers of both types of product always order full truckload quantities, the most economical scenario is already being followed.

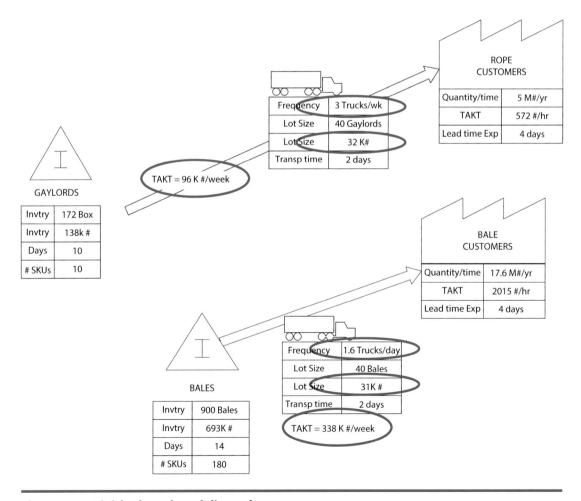

Figure 9.2 Finished product delivery frequency.

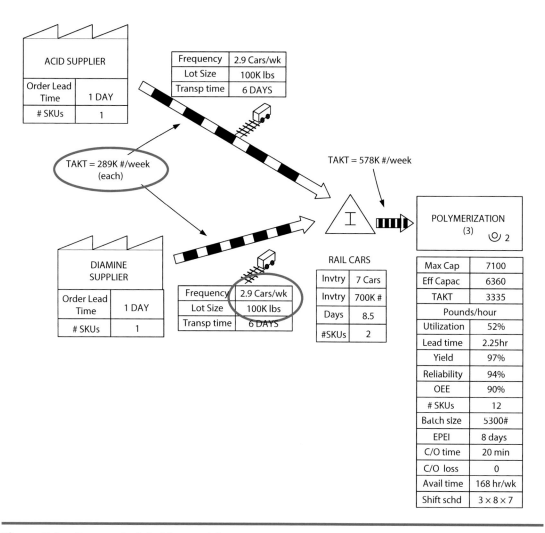

Figure 9.3   Raw material shipment frequency.

Figure 9.3 shows a similar situation; we always get diamine and adipic acid in full railcars, so smaller lot sizes are unlikely to be economically feasible.

In each case, the railcar holds 100,000 lb, and with a Takt of 289,000 lb/week, we need 2.9 railcars of each material each week.

Inventory Days of Supply

Days of supply is a very useful data box parameter; it allows you to decide if inventories are appropriately sized to satisfy the reason for having inventory at that spot in the process. Understanding where inventories are not at appropriate levels is critical because excess inventory is often the biggest waste in the process.

Days of supply can be calculated for the quantity stored and the Takt using Little's Law:

$$Days\ of\ supply = Inventory/Throughput$$

In a well-synchronized process, throughput is equal to Takt, so Little's Law becomes:

$$Days \ of \ supply = Inventory/Takt$$

Therefore, for the bale inventory shown in Figure 9.2, with an inventory of 693,000 lb and a Takt of 338,000 lb/week, the inventory will supply customers for 2 weeks, or 14 days.

Is 14 days reasonable for bale inventory? There are mathematical techniques for determining inventory requirements based on lot sizes, lead times, and variabilities, explained in Appendix E, and they should be used to analyze every inventory. However, there is also a rule of thumb, a very rough approximation, which can be used quickly to decide if an inventory quantity is in the right ballpark. The principle is that the inventory should be about 70% to 100% of the production cycle (the EPEI) of the process feeding it, based on the following logic. The 120 products baled on the four balers are each baled once every 7 days, so for each product the cycle stock will start at 7 days and drop to zero by the time its turn comes up again, so its average inventory would be 50% of that—3½ days. But there is generally some safety stock needed, which will typically raise the inventory to 70% to 100%. At any point in time, the products are all at different places in their inventory cycle, so the total inventory of all products will be reasonably consistent at 70% to 100% of the campaign cycle time. That says that the baled inventory should be 5 to 7 days, so 14 days seems excessive.

The Gaylord inventory is 138,000 lb, and with a Takt of 96,000 lb/week, Little's Law gives us 1.44 weeks, or about 10 days. This also seems to be excessive, based on the 7-day production cycle of the steam anneal machines.

Moving back through the process to the cutter box inventory, we see an inventory of 300,000 lb to supply a Takt of 345,000 lb/week, for about 6 days of supply, and that appears to be appropriate, at least at this level of analysis.

Filament tub inventory (Figure 9.1) is 640,000 lb, giving 9 days at a Takt of 493,000 lb/week. That appears to be on the high side based on the fiber spinning cycle of 7 days. The flake inventory in the silos, on the other hand, seems appropriate at 7.5 days against an 8-day polymer cycle.

The on-site inventory of adipic acid and diamine is 7 railcars, 700,000 lb, and 8.5 days at the 578,000 lb/week consumption. The 2.9 cars per week delivery frequency means that a railcar of each arrives approximately every 2.4 days. With a very reliable supplier and low variability in transportation time, 8.5 days of inventory looks very high.

We will revisit these inventories and perform the more accurate calculations when we analyze the map for opportunities to remove waste and improve flow in Chapter 12.

Summary

We have now completely developed the material flow portion of the VSM as shown earlier in Figure 2.8. The information flow portion is equally important, so we'll tackle that next.

Chapter 10

Mapping the Information Flow

Why Map Information Flow?

The purpose of the information flow on the VSM is to illustrate the impact that information processing has on physical material processing, and to show how information is managed and transformed to create schedules for the manufacturing floor. The information flow is there to explain how we know what customers need and then how we control our operations to satisfy those needs.

The information flow portion of the VSM should show all information received from customers, including specific orders and forecasts, information from manufacturing management about capacity and any other plans for that capacity, and information about inventories and their current status, so that we can see what information is available on which to base production decisions. It should show the output of those decisions and how they affect production scheduling and raw material ordering.

Including this on the VSM is critical because the material flow tells only half of the story. The material flow shows the physical process steps and highlights wastes, barriers, and problems, but we need the information flow to see the causes of some of those problems and wastes. Material flow and information flow are integrated, intertwined processes even if people don't think of them that way.

Fiber Manufacturing Information Flow

The information flow starts with the customer. Fiber customers call the Riverside Plant Customer Service Representative (CSR) assigned to their account to place an order. The CSR checks the current inventory to see if the material is available.

If it is, the CSR releases the material for shipment and confirms the shipment with the customer. These communications are all shown in the information flow (Figure 10.1).

Figure 10.2 shows the next steps in the information processing. Based on input from customers and historical trends, CSRs prepare sales forecasts for the next quarter. All CSR forecasts go into a Demand Management process for the entire business. Based on this input, other inputs, and historical trends, a forecast for the next month is prepared. This forecast is input into the monthly Sales and Operations Planning (S&OP) process.

The plant Production Supervisor prepares a Capacity Forecast for the coming month, incorporating any planned shutdowns, clean-ups, and test production runs. The S&OP process reconciles the demand forecast with available capacity and generates a Production Schedule for the coming month.

Based on the Production Schedule, the Production Supervisor issues a daily production plan for the next seven days each week. The daily production plan may be updated each day as needed to reflect the previous day's performance.

One very significant problem can already be seen from this portion of the combined VSM. The production schedules are created from forecasts, seasonal and market trends, and available plant capacity, but with no input on current inventory status. This is a huge gap in the planning process; if forecasts are high for several months, inventory will grow because there is no feedback or adjustment. This highlights the problem with a push replenishment system, which this certainly is.

There may be an informal undocumented schedule adjustment process, executed when someone notices the high inventory, but it is not shown on the initial information flow map because it is not included in any official scheduling process description documents. If this informal process exists, and can be verified, it will be shown on later versions of the VSM information flow.

If, on the other hand, the forecasts are low, it will become apparent because of the resulting stockouts.

Figure 10.3 shows that raw material orders are created in advance for an entire month and communicated to suppliers. This causes railcars to be released based on the monthly schedule; if the production forecast is high, it will cause an excessive number of railcars to be stored at the site. This is very likely the reason we have an average of 8.5 days of raw materials, where 3 to 4 days would probably be sufficient.

Capacity Constraint Resources

We sometimes see Capacity Constraint Resources (CCRs) in a process. These are manufacturing resources that cannot handle the demand put on them, i.e., they cannot make Takt, not because of any physical limitations inherent in the equipment but because of the way they are scheduled. If the flow through closely

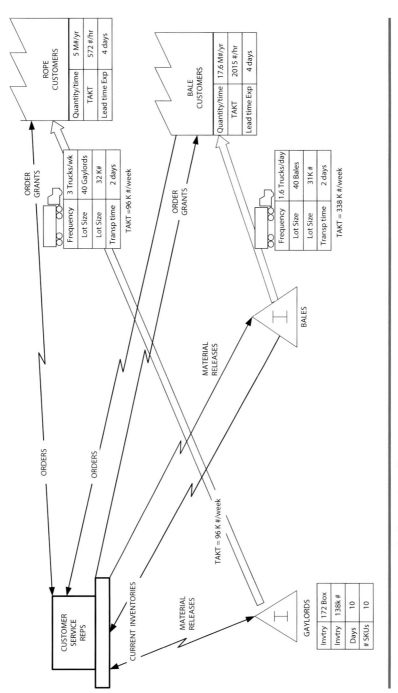

Figure 10.1 Customer information flow.

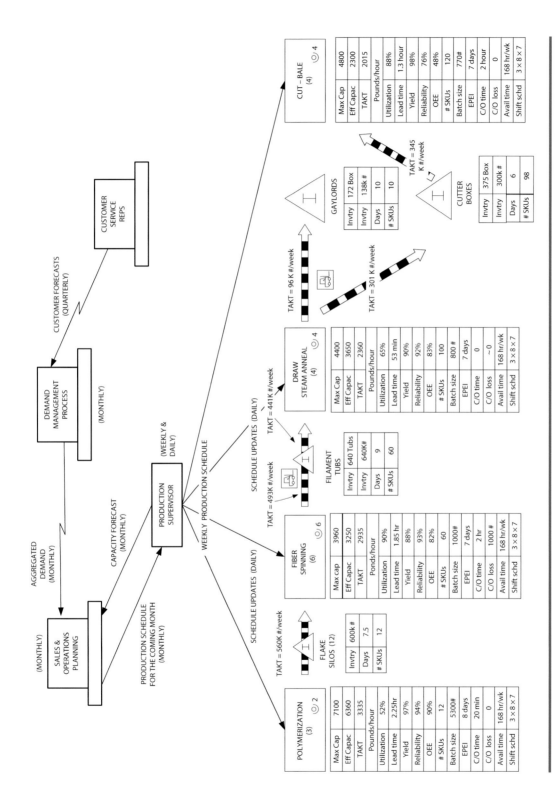

Figure 10.2 Production scheduling information flow.

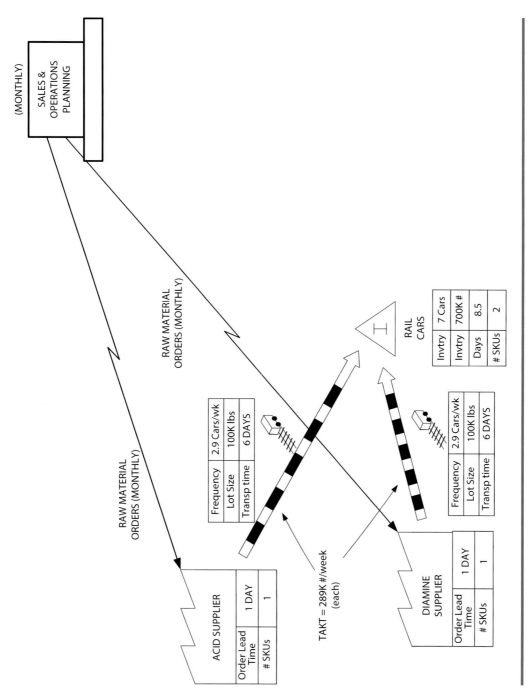

Figure 10.3 Raw material order information flow.

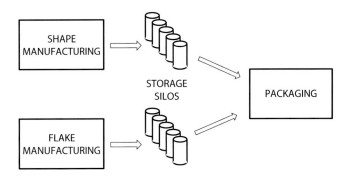

Figure 10.4 High-level view of a cereal plant.

coupled assets in a manufacturing operation is not well coordinated, if the flow through any piece of equipment is not synchronized with upstream and down-stream process steps, flow discontinuities and wasted asset time can result. *Synchronous Manufacturing* (Umble and Srikanth, 1990) defines CCRs as follows:

> **Capacity Constraint Resource**—Any resource which, if not properly scheduled and managed, is likely to cause the actual flow of product through the plant to deviate from the planned product flow.

As an example of a CCR, consider the case of a cereal plant that manufactures two families of cereal, one formed into thick shapes like stars and circles, and one formed into relatively flat flakes of various shapes. The plant can be divided into three major areas (Figure 10.4): shape manufacturing, flake manufacturing, and packaging, which includes bagging, boxing, cartoning, and palletizing.

From the data boxes contained on a more detailed map, packaging has a utilization of only 75%, even though it takes the full output of both cereal production lines. However, in real life the storage silos often became full and forced a production line to go down. Analysis revealed that although the packaging area appeared to have excess capacity, it was being scheduled with no coordination or synchronization with either production area, so it became a constraint.

The point of all this is that the combination of information flow with material flow on the VSM can often lead to an understanding of any CCRs that exist in our process.

Additional Information Mapping Tools

The information flow portion of the map can often highlight problems that actual communications cause for material flow, but not always the root causes. It is frequently necessary to dig deeper, and one of the most effective tools for a deeper dive is the cross-functional process mapping technique proposed in *Improving Performance—How to Manage the White Spaces on the Organization Chart* (Rummler and Brache, 1995). You may be familiar with these maps by their more common name—swim lane charts. They are a powerful way to break down all the interactions and hand-offs that can occur with the information processing, and get a deeper understanding of the root causes of material flow problems.

Summary

The purpose of the information flow on the VSM is to clarify how our processing of this information impacts the processing of the physical material, how customer needs are communicated and processed to manage our operations to satisfy those needs. Showing this on the VSM is critical so the interactions between information transformation and material transformation become visible. It is only through an integrated analysis of both of these flows that we can fully understand why the process is performing the way that it is and how to improve it.

In the early days of our Continuous Flow Manufacturing initiative at DuPont, our Line Analysis mapping did not include an information flow component, so we supplemented them with swim lane maps.

Chapter 11

Developing the Timeline

Timeline Principles

The third component to be shown on a VSM, after the material flow and the information flow have been charted, is the timeline, a square wave at the bottom of the VSM. Its purpose is to contrast non-value-add time with value-add time and to dramatize that for most of the time that material spends on a plant, nothing useful is happening to it. Because most of the waste in any process adds time to the process, the timeline is excellent at depicting waste. In addition, because long manufacturing cycle times (the total time that any unit of material spends on the plant) cause a lack of responsiveness and flexibility, the timeline highlights this as well.

The general convention is that the upper side of the square wave is the non-value-add time and the lower portion is the value-add time, although this is not completely standardized; some mappers do the opposite. That doesn't really matter much as long as the scheme being followed is consistent and completely understood.

The timeline should be very easy to create; all of the information needed should be in the material flow data boxes. However, the ease of generating it should not cause you to trivialize its value. The benefits of the timeline include the following:

- It highlights time wasted in the process.
- It therefore highlights other wastes in the process, such as inventory.
- It illustrates the agility and responsiveness (or lack thereof) of this process as a link in the supply chain.
- It provides a highly visible benchmark to gauge improvement.
- It provides a convincing tool for communicating with managers who have not taken the time to analyze the VSM.

Fiber VSM Timeline

To add the timeline to the synthetic fiber VSM we have been working on, we'll start with railcar storage (Figure 11.1). This storage averages 8.5 days, and is non-value-adding, so we show it as the top of the square wave. Next we have polymerization, where it takes 15 minutes to charge the raw materials into the reactor, 90 minutes for the polymerization reaction to occur, and 30 minutes to discharge the material and extrude the flake. Of this, only the 90-minute reaction is value-adding, so we show that at the bottom of the square wave. Some would argue that the 30 minutes of extrusion and flake cutting is transforming the material in a value-adding process. It is true that we are transforming the material, but that doesn't make it value-adding; this could be viewed as "packaging" the material to make it easier to store in the silos. If we directly coupled a polymer reactor to a spinning machine, we wouldn't need the extrusion step and the customer would never know the difference. On that basis, the 45 minutes spent charging and discharging is non-value-adding and could be shown as separate excursions on the timeline, or be added in with the non-value-add times of the upstream and downstream inventories, but the 15-minute charging would get lost in the noise in the 8.5-day inventory, so we can ignore it for the purpose of the timeline. A general principle of timelines is that you don't need to aim for precision. The non-value-add times are generally

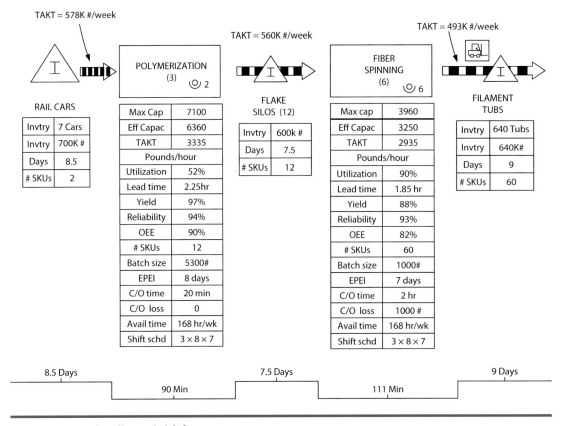

Figure 11.1   **Timeline—initial process steps.**

measured in days, while the value-add times are measured in minutes, so the important messages will be obvious even without a high degree of precision.

The flake is stored in a silo for an average of 7.5 days of non-value-add time, so that is shown at the top of the wave. It should be emphasized that even though we refer to these inventories as non-value-adding, that does not mean that they are not necessary. We need them (or some portion of them) to ensure smooth flow through our process. We highlight them as non-value-adding to focus attention on them so that we can (1) make sure that they are appropriately sized to compensate for current process performance, and (2) to understand what changes in process performance can be made so that we can reduce the amounts stored.

Spinning is next. It takes 111 minutes to spin the 1000 lb of fiber to fill a tub, all value-add time, so that's what we'll show. Some practitioners would argue that we should not consider a full tub as the unit of production, but something much smaller, say, one piece of flake. Because it takes an element of flake only a minute or two to get through the spinning process and into the tub, a minute is what should be shown. It is not worth spending much time arguing these philosophical questions unless the answers will have an impact on the message being conveyed, and in this case, not so much. So, we'll stay with the 111 minutes.

The filament tubs sit in inventory for 9 days, clearly non-value-add, thus shown at the top of the square wave.

Figure 11.2 shows the rest of the timeline, moving out toward the customer. The draw–anneal machines are next. The steam anneal machines have a value-add time of 53 minutes, while the hot roll machine has 98 minutes. Some recommend using the longer time in these situations, while I think it is more appropriate to use a weighted average. We know that 10% of the total flow goes through the hot roll machine, and 90% through the steam annealer. So the weighted average lead time is:

$$Lead\ Time = 10\% \times 98\ min + 90\% \times 53\ min = 58\ min$$

For the same reasons, we'll use a weighted average for the inventory in Gaylords, 10 days, and in cutter boxes, 6 days.

Figure 11.3 shows the entire process and complete timeline. The bottom of the wave, the value-add time, adds up to 339 minutes, or 5.6 hours. The non-value-add time, the top of the wave, totals 46 days. Manufacturing Cycle Efficiency (MCE) is a metric that takes value-add time as a percentage of the total time, and here we have an MCE of 0.5%. This may seem like an incredibly low number, but it is actually quite common that MCEs are below 1%. As you begin to reduce time waste in your process, MCE provides a good benchmark against which to demonstrate progress.

Purists might say that the timeline is not complete, that we have neglected the time wasted in the transportation steps, the time the lift trucks spend moving tubs, cutter boxes, Gaylords, and bales around. And they would be correct.

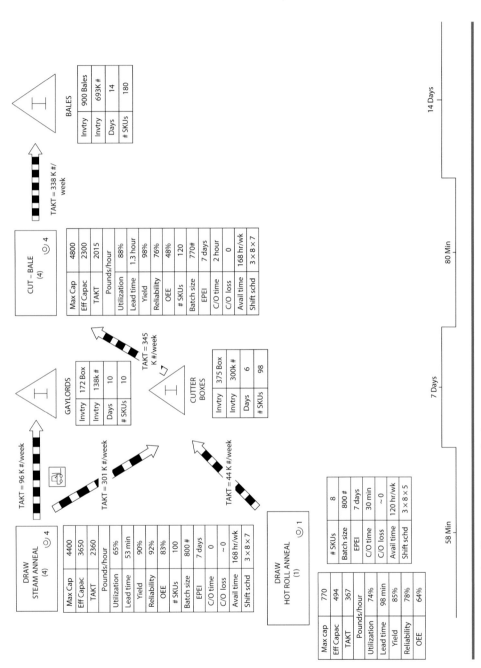

Figure 11.2 Timeline—customer facing process steps.

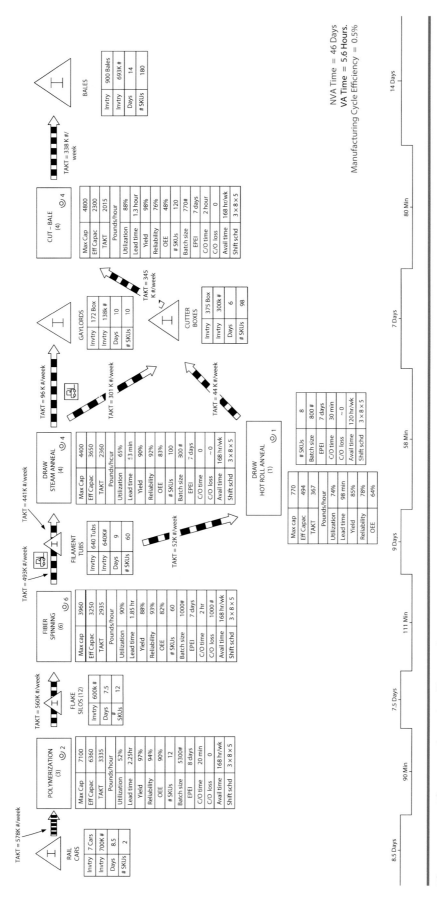

Figure 11.3 The complete timeline and MCE.

However, these time wastes are in increments of 5 to 10 minutes, and get lost in the noise when they are moving into or out of an inventory measured in 7 to 14 days. As we said earlier, the goal of a VSM is not to be completely precise, but to highlight waste and opportunities.

Cash Flow Cycle Time

Many companies measure Cash Flow Cycle Time (CFCT), the time that elapses from the time they pay for raw materials to the time they receive payment from customers for finished products, and strive to reduce it. CFCT is calculated using this formula:

CFCT = Manufacturing Cycle Time + Accounts Receivable – Accounts Payable

Accounts Receivable and Accounts Payable are based on negotiations with customers and suppliers, so Manufacturing Cycle Time is the thing most directly under your control. If you want to reduce CFCT, Manufacturing Cycle Time is the place to start, and the VSM timeline will show you where to focus.

Summary

The timeline should be very easy to add to the VSM if you have done a thorough job on the data boxes, and it can be a very useful high-level indicator of how much on-plant time is actually waste and specifically where those wastes are. If we want to shorten Manufacturing Cycle Time to make the process more responsive to changes in customer demand, the timeline will show where to focus our efforts.

We have now completed the Riverside Plant fiber manufacturing VSM. So? Well, now we can put the VSM to good use by analyzing it to see where we can reduce waste and improve process flow. Then we'll scope and prioritize those improvements and document the possibilities on future state VSMs.

Chapter 12

Finding the Waste—
Analyzing the Map

Now that we have developed the Riverside Plant fiber VSM, it can provide five useful and valuable functions:

1. Bringing a higher level of clarity on how the process is currently performing and why
2. Showing all the major wastes in the process
3. Highlighting areas where improvement should be made
4. Providing a roadmap for the implementation of Lean concepts to improve flow and process performance
5. Providing a template for documenting the ideal future state and nearer term steps toward the final future state

If the VSM did only the first of these, it would be well worth the time and effort that went into its creation. However, it can do so much more that it is a mistake to stop at that point.

Many references recommend that after completing the current state VSM, the next step is to create a future state VSM. They suggest that you should decide where within the process to work toward continuous flow and then develop a pull flow strategy. Although that will sometimes work, there are frequently enough process issues, variabilities, and instabilities, that going directly to pull or to any final future state can be difficult. Achieving the future state can be much more feasible if the primary problems with current performance are identified and then resolved, following a multi-generation future state plan.

Therefore, the creation of a future state VSM should begin with a careful analysis of the current state map to see what it can tell you about your process, its current performance, and opportunities to reduce waste. The more challenging future state goals are much easier to reach if the process has first been stabilized and improved. Thus, the move to a defined future state should be more than going to pull replenishment and continuous flow; it should include correction of all significant performance detractors present in the current state, which can be found through a careful analysis of the current state VSM.

General Impressions from the Current State

Looking at the portion of our current state VSM shown in Figure 12.1, we can make some high-level judgments on what we see.

- Capacity is not a significant concern. There are no bottlenecks or near-bottlenecks; utilizations are all at or below 90%. In fact, as noted earlier, polymerization utilization is 52%, suggesting elimination of one reactor. In addition, the 65% utilization of the steam anneal machines offers the possibility of shutting down the hot roll machine.
- Inventories appear to be excessive at several places: raw material railcars, filament tubs, and finished products in Gaylords and in bales. We can't be sure until we've done a mathematical analysis based on lead times, production quantities, and demand variability, but the VSM is raising flags.
- Each major process step is scheduled based on demand forecasts without regard to current inventory status. This push scheduling strategy is likely one of the causes for the high inventories we see.
- Overall yield is an issue. The Takt numbers on the VSM indicate that the combined customer demand for baled staple fiber and rope in Gaylords is 434,000 lb/week. We consume 578,000 lb of raw materials per week to produce that output, so only 75% of the material going in comes out as first grade product. We waste 25% of everything we purchase!

Inventory Opportunities

Inventories seem too high at several places. Figure 12.2 tabulates the inventory that the rule of thumb discussed in Chapter 9 would suggest, the current inventory, and the difference. It should be emphasized that a difference doesn't necessarily mean that the current amount is inappropriate, just that it warrants a more accurate analysis.

A few comments are in order:

- The biggest opportunities appear to be at the input and output ends of the process, in raw materials and the finished products. This is not uncommon.
- Flake appears to be appropriately sized based on the 8-day polymerization campaign cycle. But the 8-day cycle may be far too long, and a reduction would enable a corresponding reduction in flake inventory.

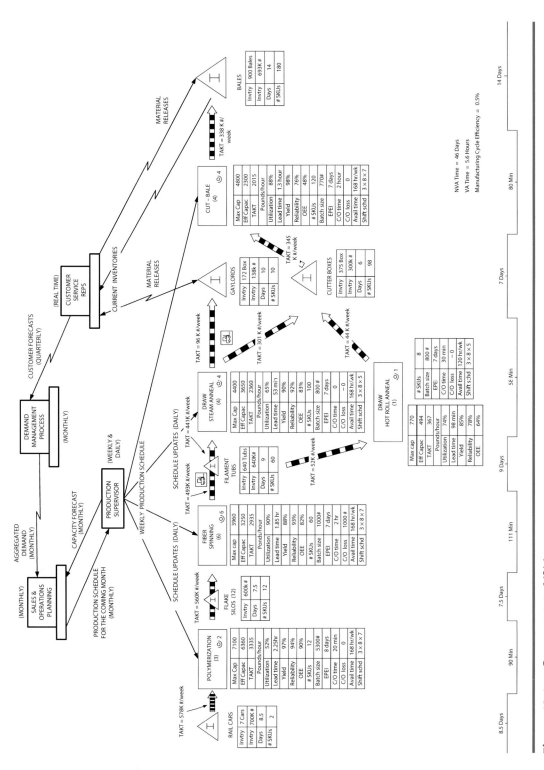

Figure 12.1   Current state VSM.

MATERIAL	INVENTORY LOCATION	RULE OF THUMB	ACTUAL	DIFFERENCE
Diamine, Adipic Acid	Rail Cars	5 days	8.5 days	3.5 days
Flake	Silos	5–8 days	7.5 days	0
Filament	Tub storage	5–7 days	9 days	2–4 days
Rope	Gaylord storage	5–7days	10 days	3–5 days
Rope	Cutter Box Storage	5–7 days	6 days	0
Cut Fiber	Bale Inventory	5–7days	14 days	7–9 days

Figure 12.2 High spot inventory opportunities.

■ Rope in cutter boxes appears to be appropriately sized.

To repeat a key point, this just a set of approximations based on a rule of thumb that has proven to be useful in past applications but doesn't always tell the full story. A more mathematically rigorous analysis must be done to determine which apparent opportunities are real.

Baler Reliability

It can be seen from the baler data box (Figure 12.3) that reliability is poor at 76%, which drives OEE down to 48%. Even with this low OEE, utilization is still an acceptable 88%, so this is not a throughput issue on average, but it could be a serious throughput problem during downtimes. It also implies high maintenance cost to keep the balers running. It should be addressed because reliable equipment that's available when you need it is a basic requirement of any Lean production system.

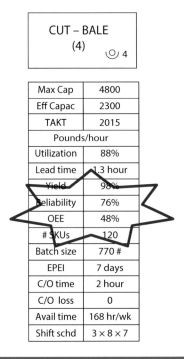

CUT – BALE (4) ☉ 4	
Max Cap	4800
Eff Capac	2300
TAKT	2015
Pounds/hour	
Utilization	88%
Lead time	1.3 hour
Yield	98%
Reliability	76%
OEE	48%
# SKUs	120
Batch size	770 #
EPEI	7 days
C/O time	2 hour
C/O loss	0
Avail time	168 hr/wk
Shift schd	3 × 8 × 7

Figure 12.3 Baler reliability.

Spinning Yield

The Spinning data box in Figure 12.4 shows a yield of 88%. From the Takt numbers we see that 67,000 lb are lost each week in spinning. That's almost half of the 144,000-lb yield loss across the entire process, and nearly all of it occurs during changeovers. Two potential solutions:

1. Find a way to make the changeover more quickly.
2. Find a way to throttle flake feedback further during a changeover.

Long Campaign Cycles (EPEIs)

The 8-day EPEI in polymerization looks far too long. Figure 12.4 shows that changeovers can be done quickly, 20 minutes, without any losses or out-of-pocket cost. Further, there are only 12 SKUs to produce on the three reactors. A much shorter campaign cycle, say 2 or 3 days, would allow a dramatic reduction in flake inventory, and wouldn't increase utilization significantly.

Utilization would still be low enough that one reactor could be taken offline with the remaining two handling the full Takt.

The draw–steam anneal machines provide another place to look at EPEI reduction. With zero changeover time and zero changeover losses, there is no reason why it has to be 7 days, and could be much less, which would allow reductions in Gaylord and cutter box inventories.

Hot Roll Draw–Anneal

The hot roll machine performs very poorly. The data box in Figure 12.4 shows 78% reliability and 85% yield, each much lower than the steam anneal machines. We could suggest aggressive programs to improve both of these parameters. However, a better alternative would be to qualify the products that currently must be made on the hot roll machine to run on the steam anneal machines, and shut the hot roll machine down. Benefits would accrue in yield, energy cost, maintenance cost, and operating labor cost.

Uncoordinated Scheduling

The information flow in Figure 12.1 shows that each major process step is scheduled independently of the others, giving evidence that the process is viewed as a collection of uncoordinated pieces rather than as an integrated system. Even worse, current inventory levels are not considered in the production scheduling

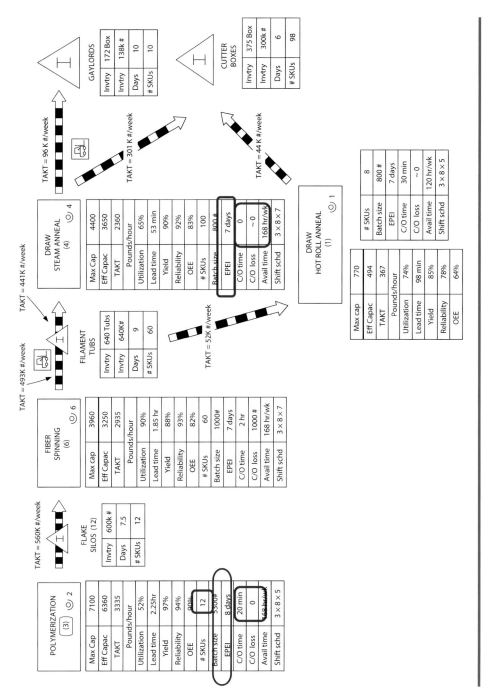

Figure 12.4 Candidates for lower EPEI.

process; schedules are based solely on forecasts. This is a pure "push" system, and thus it is no surprise that inventories are excessive.

A "pull" replenishment system is needed. In a pull system, equipment is not centrally scheduled based on forecasts. What is produced is based on the current contents of the downstream inventory. In a well-designed pull system, there should be a general scheduling strategy to balance flow and manage the product sequence, but what actually is produced within that strategy is what has been consumed from the downstream inventory during the past cycle. Thus inventories can't exceed their targets and overproduction is avoided. And scheduling gets simpler and timelier. Forecasts are still needed for long-term planning, to ensure that we have the assets and personnel needed, and to set inventory targets. However, what we make today should be based on what was consumed yesterday.

Pull replenishment systems are described more fully in Appendix D.

Capturing Potential Opportunities

Putting a starburst symbol around each feature of the current state VSM that may be an opportunity worth pursuing is a common way to highlight all of the wastes and flow problems the map shows. Figure 12.5 illustrates this for the fiber process.

Now that we have looked the VSM over for likely opportunities, the next steps are to:

1. Develop each idea to verify that it is a true opportunity.
2. Scope the actions necessary to capture the opportunity.
3. Estimate the benefit of each one.
4. Prioritize them.
5. Decide on the most appropriate sequence of implementation.
6. Develop a future state VSM, or a series of sequential future state VSMs to predict the performance we can expect after successful implementation.

Chapter 13 will take you through the first three of these, Chapter 14 the next two, and Chapter 15 the development of the future state VSMs.

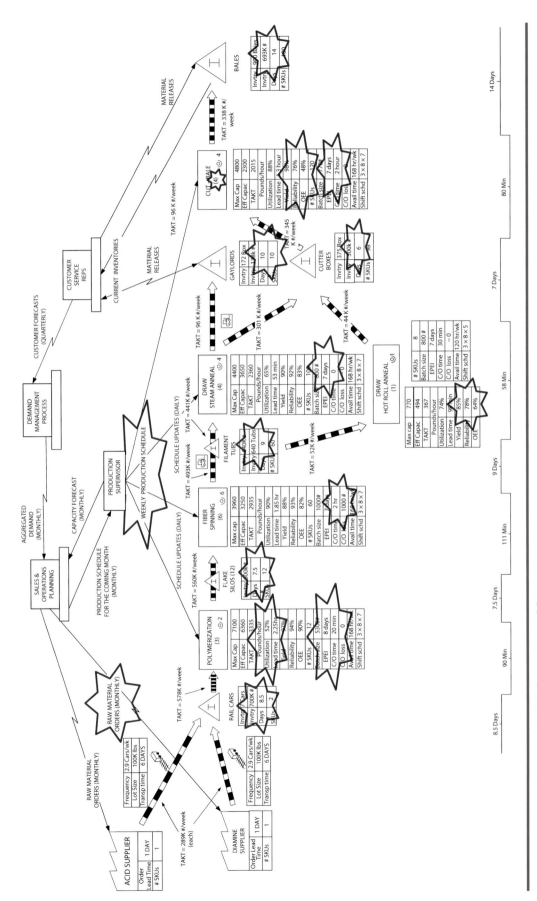

Figure 12.5 Starbursts to highlight opportunities.

Chapter 13

Scoping the Opportunities

Now that we have analyzed the current state VSM, found the most substantial wastes, identified other possible improvements, and highlighted the opportunities with starbursts on the VSM, it is time to further develop each idea to understand the benefit it is likely to give and any potential barriers to implementation. When you are scoping out all the likely opportunities from a VSM, it is important to use a reasonably consistent format so that opportunities can be compared and prioritized based on benefit and on likelihood of successful implementation. And the benefits should include any Lean-focused benefits in waste reduction and flow improvement as well as business benefits in revenue enhancement and cost reduction.

Inventory Opportunities

1. Raw Material Inventory in Rail Cars Is Too High

Figure 13.1 shows the raw material portion of the VSM. Raw material inventories appear to be high, probably because orders are placed monthly in advance based on a forecast. The calculations described in Appendix E indicate that if orders are placed as needed based on actual consumption (i.e., pull), an inventory of 3 to 4 railcars gives sufficient cycle stock as well as the safety stock needed to protect against the variability in consumption. That suggests a reduction of 3.5 railcars, 350,000 lb.

> **Scope of improvement:** Reduce rail car inventory to that needed based on delivery frequency and lead time; use pull concepts to schedule replenishment.
> **Benefit:** One-time savings of $240,000; ongoing annual savings of $64,000.
> **Difficulties:** Nothing significant.
> **Implementation plan:** Modify the process for communicating with the suppliers, to base re-ordering on a kanban-type trigger point rather than a monthly forecast.
> **Timing:** Immediate, within a month.

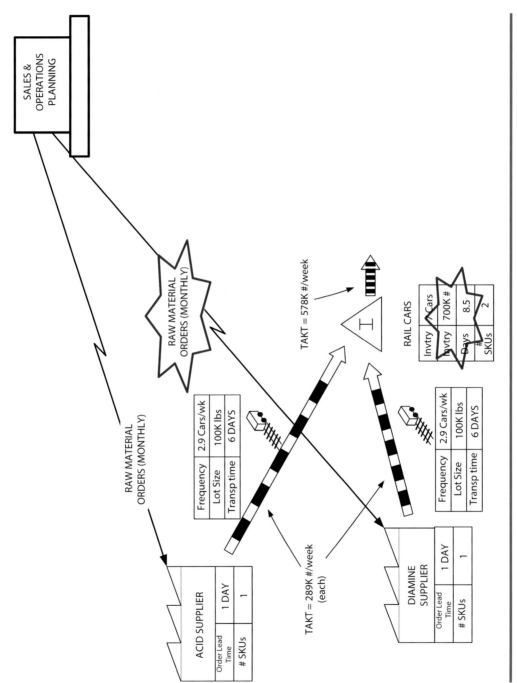

Figure 13.1 Raw material portion of the current state VSM.

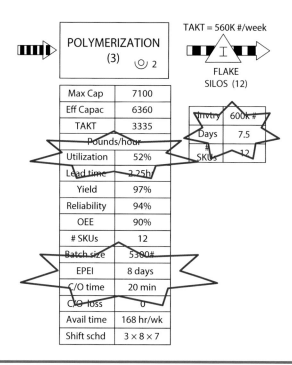

Figure 13.2 Polymerization and flake opportunities.

2. Flake Inventory Is High

Figure 13.2 suggests that the current flake inventory (7.5 days) is in the right ballpark for the current 8-day polymerization campaign cycle. However, there is no reason for a cycle that long; changeovers are very quick, and there are no changeover losses. A 2-day cycle appears very feasible, which will allow a significant inventory reduction.

Scope of improvement: Reduce flake silo inventory by reducing polymerization EPEI to 2 days using product wheel scheduling; use pull concepts to schedule replenishment.

Benefit: One-time savings of $365,000; ongoing annual savings of $98,000.

Difficulties: Nothing significant.

Implementation plan: Design and implement a product wheel scheduling strategy. (Product wheels are described in Appendix F.)

Timing: Nothing immediately—inventory is close to the required amount for current performance and EPEI. Full benefit will come when product wheels are implemented, which is best done after cellular flow is operational.

3. Filament Inventory Is High

Scope of improvement: Reduce filament tub inventory to the proper amount to manage current performance, 6.3 days, 450,000 lb.; use pull concepts to schedule replenishment.

Benefit: One-time savings of $200,000; ongoing annual savings of $56,000.

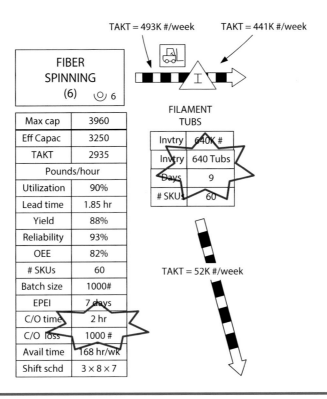

Figure 13.3   Spinning and filament opportunities.

> **Difficulties:** None.
>
> **Implementation plan:** Implement pull replenishment. Inventory will naturally ramp down to the required level if you stop producing any high inventory SKUs until their inventory drops to the appropriate level.
>
> **Timing:** Now.

4. Rope Finished Product Inventory in Gaylords Is Too High

> **Scope of improvement:** Ramp inventories down to the levels required to support current performance; use pull concepts to schedule replenishment.
>
> **Benefit:** One-time savings of $60,000; ongoing annual savings of $16,000 through reduced inventory.
>
> **Difficulties:** Nothing significant—simply put new inventory targets in place.
>
> **Implementation plan:** Implement pull replenishment. Inventory will naturally ramp down to the required level if you stop producing any high inventory SKUs until their inventory drops to the appropriate level.
>
> **Timing:** This could be done now, but is better done after the cellular flow plan has been established. At that point, it is likely that one draw–anneal machine will be dedicated to rope products. (See Opportunity 14.)

5. Right Size the Cutter Box Inventory

The inventory calculations show that we actually have slightly less cutter box inventory than needed for desired performance; when mathematical techniques

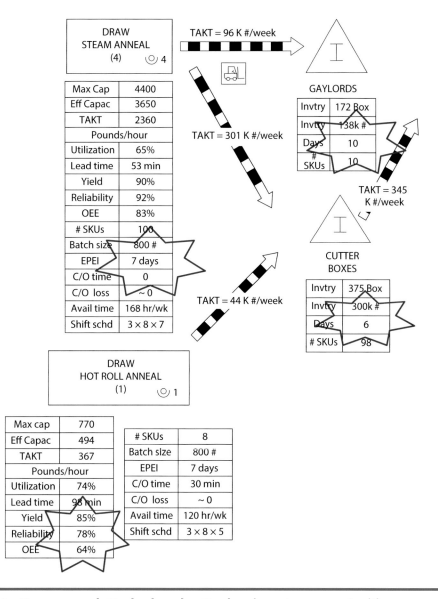

Figure 13.4 Draw–anneal, Gaylord, and cutter box inventory opportunities.

are brought into inventory management, it sometimes reveals that inventories are actually lower than they should be. If they were significantly lower than needed, it would have been obvious from the frequent stockouts. In this case, the shortfall is minor; the calculations show that 6.3 days is required to meet the 98% service level target, compared to the 6 days currently held. So the recommended increase is slight but should be done. This will increase cost, and from a Lean perspective will increase waste, but that is better than shorting customers. While Lean abhors waste, everything done should be based on customer value and service.

Scope of improvement: Increase the cutter box inventory to the recommended level; use pull concepts to schedule replenishment.

Benefit: Improved delivery performance for our customers.

Difficulties: Nothing significant—there will be a one-time cost of $12,000 plus an ongoing annual cost of $4,000 through right-sized inventory.

Implementation plan: Recalculate inventory requirements, and then implement pull replenishment to ramp up to them.

Timing: Should be done as soon as practical to prevent an unacceptable frequency of stockouts.

6. Bale Finished Product Inventory Is Too High

There are two things that can be done to reduce bale inventory (Figure 13.5): (1) reduce bale inventory to the level needed based on current Cut–Bale frequency and lead time, and (2) further reduce it by reducing the overall production cycle (EPEI).

Scope of improvement: Reduce baled inventory to that needed based on performance, 300,000 lb, 6.3 days. Then implement product wheels to shorten the overall production cycle; size the inventory in accordance with the shorter cycles; use pull concepts to schedule replenishment.

Benefit: The first step will generate a one-time savings of $525,000 and an ongoing annual savings of $142,000.

Difficulties: We should implement an improved scheduling methodology such as product wheels on the balers to reduce campaign cycles. However, with utilization currently at 88%, reliability will have to be improved for this to be feasible. Reducing changeover time will also help. Therefore,

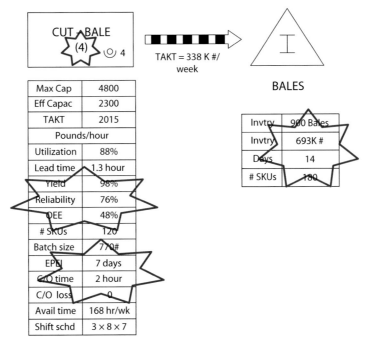

CUT–BALE	
(4) ◎ 4	

TAKT = 338 K #/week

BALES

Max Cap	4800
Eff Capac	2300
TAKT	2015
Pounds/hour	
Utilization	88%
Lead time	1.3 hour
Yield	98%
Reliability	76%
OEE	48%
# SKUs	120
Batch size	770#
EPEI	7 days
C/O time	2 hour
C/O loss	0
Avail time	168 hr/wk
Shift schd	3 × 8 × 7

Invtry	900 Bales
Invtry	693K #
Days	14
# SKUs	180

Figure 13.5 Cut–Bale and bale inventory opportunities.

the reduction in campaign cycles must wait until Opportunities 8 and 9 have been accomplished.

To get the benefit listed here we simply need to calculate proper inventory levels and then ramp down to them.

Implementation plan: Implement pull replenishment. Inventory will naturally ramp down to the required level if you stop producing any high inventory SKUs until their inventory drops to the appropriate level.

Timing: Can be done as soon as practical.

Equipment Opportunities

7. Spinning Changeover Losses Are High; Spinning Utilization Is High

The spinning data box (Figure 13.3) highlights that spinning changeovers take 2 hours and waste 1000 lb of material. This amounts to the loss of 60,000 lb/week, at a cost of more than $3 million per year. This is about half of the yield loss across the entire process, and probably the largest single waste. However, reducing it will not be easy. The best approach will be to reduce the changeover time using SMED (Single Minute Exchange of Dies) change-over reduction techniques. (The SMED process is described in Appendix B.) We should also find a way to further reduce polymer flow during a changeover, but that is less likely to be successful.

While a spinning utilization of 90% is not a serious concern, it is the highest utilization in the entire production process, and therefore the most likely to become a bottleneck if product demand grows. Anything done to reduce utilization will be a benefit, and a changeover time reduction will accomplish that.

Scope of improvement: Reduce changeover losses by reducing changeover time, using SMED techniques. The goal is to cut the 2-hour changeover time to 1 hour. In applying SMED, look at labor loading the changeover to take time out.

Benefit: Ongoing annual savings of $1.6 million through reduced changeover losses. OEE will increase from 82% to 87.4%. Utilization will drop from 90% to 85% through reduction of changeover time, which gives more breathing room.

Difficulties: It is not a given that we can cut losses in half. Difficulty is at least moderate.

Implementation plan: Stage a SMED Kaizen event. To achieve the benefits targeted, it may take more than one crack at it. So the plan is to hold a SMED Kaizen event, take changeover time down as much as possible, stabilize the operation at that level, and then stage another Kaizen event to generate another round of improvement ideas. Repeat as necessary.

Timing: SMED can be done as soon as team can be assembled and preparations made.

8. Baler Reliability Is Poor

One of the starbursts in Figure 13.5 flags baler reliability (76%) as a problem. This is a waste of valuable asset time, and incurs maintenance costs that are clearly waste.

Scope of improvement: Implement better maintenance practices to improve baler reliability above the current 76%. The target is 82%; this would increase OEE from 48% to 52%, and drop utilization from 88% to 81%.

Benefit: Reduces maintenance cost and lost time; provides more tolerance for significant upsets. More importantly, this is one step (along with changeover time reduction via SMED, Opportunity 9) toward reducing EPEI (Opportunity 10) and getting the additional reduction in bale inventory mentioned in Opportunity 6, or toward eliminating one baler (Opportunity 11).

Difficulties: This is a very complex piece of electro-mechanical equipment. The fibers cut to very short lengths can get into the moving mechanisms, causing jams. Therefore, reliability improvements won't be easy, but that shouldn't prevent a real effort to try.

Implementation plan: Implement better maintenance practices—perhaps TPM (Total Productive Maintenance) and autonomous maintenance; involve mechanics and mechanical engineers in a Kaizen workshop to find ways to modify the balers to reduce chronic problems and to redesign for greater reliability. Modifying airflow within the baler might allow fines to be collected in the air handling system rather than in the mechanical gears and linkages.

9. Changeover Improvement—Balers

Referring again to Figure 13.5, we see a changeover time of 2 hours. Reducing the baler changeover time will not by itself do much beyond improving OEE and reducing utilization, but it may allow for shorter campaign cycles, which will reduce baled inventory as discussed as the second part of Opportunity 6. If this opportunity and the reliability improvement in Opportunity 8 are successful, that may allow one baler to be taken offline, which would offer economic benefits in several areas.

Scope of improvement: Use SMED to reduce changeover time. Goal: 1 hour vs. the current 2 hours.

Benefit: This is one of the improvements required to either shorten baler campaign cycles and reduce inventory, or take one baler offline. OEE would improve from 48% to 61%, and to 66% when combined with the reliability improvement (Opportunity 8). Utilization would drop from 88% to 64%.

Difficulties: As described previously, this is a very complex piece of equipment, and must be thoroughly cleaned out between products. The improvements may be in improved cleaning tools and methods, or in modifying the structure to prevent fibers from getting into the mechanisms as much. The difficulty is judged to be moderate.

Implementation plan: Plan and execute a changeover reduction Kaizen Event. As discussed previously, it may take more than one iteration to reach the goal, so allow for followup Kaizen events.

Timing: Can be done now, but should be scheduled where it best fits into the priorities of all the improvements being planned.

10. Reduce the Baler Campaign Cycle (EPEI)

This is heavily interconnected with the other baler improvements. If reliability can be improved (Opportunity 8) and the changeover time reduced (Opportunity 9), then it should be straightforward to cut the 7-day EPEI in half. However, this opportunity can't be taken if the next one, to mothball one baler to reduce operating cost, is chosen.

Scope of improvement: Implement 4-day product wheels on all balers. (This assumes successful implementation of Opportunities 8 and 9, with the forecast benefits having been achieved.)

Benefit: Utilization will increase, but only to 76% so it is still very comfortable. Bale inventory will drop from 260,000 lb (5.4 days) to 150,000 lb (3.1 days). (This is in addition to the reduction achieved by right-sizing the current inventory based on current performance, Opportunity 6, and the removal of obsolete SKUs, Opportunity 17.) This will generate $150,000 in one-time savings and $40,000 ongoing annual savings.

Difficulties: Reducing campaign cycle time using a product wheel scheduling strategy will be relatively straightforward after the reliability and changeover improvements are completed. However, a product wheel this short cannot be accomplished unless the reliability and changeover improvements successfully meet targets; the optimum wheel time will have to be determined based on how successful those efforts are, and how much improvement is accomplished.

Timing: This is best done after Cellular Flow is operational, when will there be fewer products run on each baler.

Implementation plan: Shelve this idea; after reviewing this and the following opportunity, the business decided to implement a product wheel scheduling strategy, but keep the cycle at 7 days. Opportunity 11 offers far more benefit than the 4-day wheel does, and they are mutually exclusive.

11. Mothball One Baler

Scope of improvement: Take one baler out of service; reassign products to other three balers; keep the baler in place with all piping and wiring—it may be needed if something catastrophic happens to one of the operating balers or if product demand grows. An offline baler would be very useful to prototype equipment modifications to reduce fiber contamination.

Benefit: This will save operating cost in labor, energy, maintenance, etc. The savings are estimated at $600,000 per year.

Difficulties: This requires successful projects to improve baler reliability and reduce changeover time. And this will not be feasible if the decision is made to reduce baler EPEI (Opportunity 10); we can do that or this but not both. But this is the greater opportunity by an order of magnitude: $600,000 annually vs. $40,000.

Timing: This should be done after baler reliability and changeover improvements (Opportunities 8 and 9) are successful, but before cellular flow is implemented.

12. Hot Roll Draw–Anneal Reliability and Yield Wastes

The hot roll machine (Figure 13.4) has a number of large wastes associated with it: low yield, low reliability, and long changeover times when compared with the steam anneal machines. Rather than spend a lot of time and money to try to improve all these factors, it would be better to get rid of the hot roll machine and move its products to the steam anneal machines.

Scope of improvement: Take the hot roll machine out of service and dismantle it. Run the hot roll products on the steam anneal machines.

Benefit: If the products made on the hot roll machine were made on the steam anneal machines, the yield would likely be 5% higher, saving $135,000 annually in material costs. Taking the hot roll machine out of service reduces plant operating cost by $300,000 per year.

Difficulties: Process development and prove-out would have to be done on the eight products currently made on the hot roll machine so that they could successfully be made on the steam anneal machines. There doesn't appear to be any technical reason why this wouldn't be successful, so it's just a matter of finding the right process conditions and settings, and then demonstrating acceptable performance. This development is a one-time cost, while the savings are ongoing annual benefits. Moving these products will increase utilization on the steam anneal machines, from 65% to 75%, which is still very comfortable.

Implementation Plan: Develop the process conditions to run the hot roll products on the steam anneal machines; qualify the products with sample runs and customer trials. Then shut down, dismantle, and remove the hot roll machine.

Timing: This should be started as soon as process development specialists are available.

13. Mothball One Polymer Reactor

The starbursts on Figure 13.2 highlighted that the three polymer reactors have a utilization of only 52%, and will go to only 53% with the 2-day product wheel called for in Opportunity 2. Therefore, two reactors have more than enough capacity to process the full Takt.

> **Scope of improvement:** Take one polymer reactor out of service. Leave it in place, fully piped and wired for now, as demand is expected to increase over time and it may be needed in the future. This will result in the two remaining reactors being utilized 81% of the time.
>
> **Benefit:** The total annual operating cost for each reactor is $1.4 million, primarily in the energy costs to maintain the high vessel temperatures required for polymerization; taking one reactor offline will save approximately $1.2 million.
>
> **Difficulties:** None foreseen; implementation will simply be a matter of reallocating products between the two operating reactors.
>
> **Timing:** Can be done immediately.

System-Wide Opportunities

14. Implement Virtual Cellular Flow

Cellular flow and its benefits are described in Appendix C. Basically there are two steps involved: (1) group the equipment into virtual flow cells as shown in Figure 13.6, with dedicated flow paths so that material from polymer reactor 1 feeds only spinning machines 1 and 2, for example. This reduces variation in the final product and makes flow far easier to manage. (2) Break the product line up into families with similar processing characteristics and dedicate each family to a specific cell. This further simplifies flow management and generally makes changeovers faster and less wasteful. Product wheel design becomes easier because there are fewer products run on any specific piece of equipment.

> **Scope of improvement:** Implement virtual cellular flow as shown in Figure 13.6; reallocate products among the cells in a way that minimizes the product variations in any specific cell.
>
> **Benefits:** The benefits here are primarily in reduced variability and much more standardized flow patterns. There is almost always a significant economic benefit, but in ways that are difficult to predict ahead of implementation. Throughput will very likely improve due to simpler, shorter changeovers. Yield on re-start will likely improve due to simpler changeovers. The full economic benefits can't be quantified until the product families are analyzed and assigned.
>
> **Difficulties:** Nothing significant.
>
> **Implementation plan:** Perform an analysis of the plant layout to determine the most straightforward flow paths, and a thorough analysis of all product families to group the most similar products together on a cell. Document the

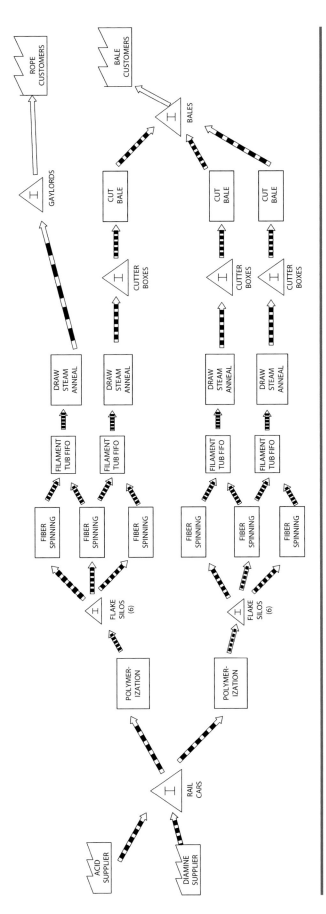

Figure 13.6   Cellular flow opportunity.

flow paths, paint flow lines on the floor, and hang signs to make the new flow patterns obvious. Document the product family assignments.

Timing: This is best done when the future equipment configuration has been established, that is, after the hot roll anneal machine, one polymer reactor, and one baler are taken out of service.

15. Filament Tub FIFOs after Implementation of Virtual Cells

Because the steam anneal machines (Figure 13.4) have zero changeover time and zero changeover losses, there is no reason for them to follow a specific sequence or to run long campaigns. Therefore, they could follow the spinning sequence and simply anneal whatever product is currently being spun. However, each steam anneal machine draws from two spinning machines, so some buffer is required. It takes less than an hour for a steam anneal machine to process one filament tub, so it would make sense for a steam anneal machine to process, say, six tubs from one spinning machine, and then process six tubs from the other. Thus, the inventory in the First In–First Out (FIFO) buffer at any time would be just a few tubs, far less than one day of inventory.

Scope of improvement: Replace planned filament tub inventory with FIFOs.

Benefit: One-time savings of $425,000 and ongoing annual savings of $114,000 through reduced inventory. These numbers are above and beyond the benefit of right-sizing the filament tub inventory to match current performance; we have already taken credit for that in Opportunity 3.

Difficulties: Depends on the difficulty of implementing virtual cells, which depends somewhat on eliminating the hot roll annealer. Neither of these is viewed as being particularly difficult.

Implementation plan: After virtual cells are in place, manage the filament tub inventories as FIFOs rather than as planned inventory.

Timing: This can be done as soon as virtual cells are in place.

16. Implement Pull Replenishment across the Value Stream

A starburst in the information flow portion of Figure 12.5 highlighted that scheduling is currently uncoordinated; each area is scheduled from a monthly forecast, without regard to current process conditions or inventories. This is a significant cause of the high inventories seen on the VSM.

Each of the inventory opportunities described previously (1–6) include implementing pull concepts to achieve the target inventory reductions. Therefore, pull will be implemented as part of the implementation plans of these inventory opportunities, and the benefits of pull replenishment have already been accounted for. They total $1.37 million in one-time savings and ongoing annual savings of $370,000. These pull benefits are tabulated in Figure 13.7. The total

| | TOTAL INVENTORY BENEFITS (K$) | | PULL BENEFITS ONLY (K$) | |
AREA	ONE-TIME	ANNUAL	ONE-TIME	ANNUAL
Rail cars	$240	$65	$240	$65
Flake	$365	$98	$365	$98
Filament	$200	$56	$200	$56
Filament FIFOs	$425	$114		
Gaylords	$60	$16	$60	$16
Cutter boxes	($12)	($4)	($12)	($4)
Bales	$525	$140	$525	$140
Low Demand SKUs	$60	$16		
TOTAL	$1,800	$500	$1,370	$370

Figure 13.7 Summary of inventory opportunity benefits.

inventory benefits, which include those coming from right-sizing the inventories based on current performance, are also shown.

17. Drop the 60 Very Low Demand SKUs

Of the 180 baled products available, only 120 are made on a regular basis; the rest are relatively obsolete products with little or no demand. But, they are listed in the catalog and inventory is carried just in case an order is received. This highlights the need for the business to have a more effective Portfolio Management process, but at the very least, these products should be dropped from the catalog, and the inventory disposed of to capture whatever value there is.

Scope of improvement: Take these very low demand SKUs out of the catalog of available products. Dispose of the current inventory of these products in whatever way generates the most revenue.

Benefit: Reduces complexity; eliminates clutter in the catalog, spreadsheets, and master data. Takes approximately $60,000 out of inventory and saves $16,000 annually in inventory carrying cost.

Difficulties: Requires management agreement: Business Manager, Marketing Manager.

Implementation plan: Select the candidate products, develop a case proposal, and present it to management. Delete these products from all master data and spreadsheets and sell off any remaining inventory (providing that doesn't cannibalize profitable products).

Timing: Can be started now.

Summary

The starbursts in Figure 12.5 dramatized that there is a lot that can be done to improve process performance and reduce waste and cost, and the scoping we have just done has reinforced that. It also made it obvious that many

NUMBER	OPPORTUNITY	ONE TIME BENEFIT (K$)	ON-GOING BENEFIT (K$)	BENEFIT SCORE	FEASIBILITY
1	Raw material inventory in rail cars is too high	$240	$65	2	5
2	Flake inventory is high	$365	$98	3	5
3	Filament inventory is high	$200	$56	2	5
4	Rope finished product inventory in gaylords is high	$60	$16	2	5
5	Right size the cutter box inventory	($12)	($4)	3	5
6	Bale finished product inventory is too high	$525	$140	3	5
7	Spinning changeover losses are high; spinning utilization is high	$0	$1,600	5	3
8	Baler reliability is poor			4	2
9	Changeover improvement — Balers			4	2
10	Reduce the baler campaign cycle	$150	$40	2	2
11	Mothball one Baler	$0	$600	4	2
12	Hot roll draw-anneal — reliability and yield wastes	$0	$435	4	4
13	Mothball one polymer reactor	$0	$1,200	5	5
14	Implement Virtual Cellular flow			4	4
15	Filament tub FIFOs after implementation of virtual cells	$425	$114	3	4
16	Implement pull replenishment across the value stream	The benefits were accounted for in opportunities 1–6			
17	Drop the 60 very low demand SKUs	$60	$16	2	4
		$2,000	$4,400	5-High	5-Easy
				3-Moderate	3-Moderate
				1-Low	1-Difficult

Figure 13.8   Summary of opportunity benefits and feasibility.

of the improvement possibilities are interconnected and interdependent, and therefore must be done in a well-coordinated sequence. We also saw that some (Opportunities 10 and 11) are mutually exclusive, that while each offers benefits they can't both be done so a choice must be made.

The benefit and the feasibility of each of the scoped opportunities are summarized in Figure 13.8. Every effort has been made to separate the benefits of those interdependent opportunities so that benefit estimates are not duplicated. Benefits are scored on a scale of 1 to 5, where 5 is high benefit and 1 is low. Ongoing benefits are given priority over one-time benefits in this scoring. The feasibility, the likelihood of successful implementation, is scored on a similar 1 to 5 scale. This scoring, especially the feasibility, is somewhat subjective, but that's all right because its primary function is to make sure that expectations are set realistically.

While all of these things are worthwhile and should be done, they can't all be done at once, so a phased approach, a multi-generational plan, must be developed to move toward the ideal future state in logical, coordinated steps. The next two chapters will walk us through a ranking and prioritization process and determination of the best sequence for implementation so that the various generations of future state VSMs can be developed.

Chapter 14

Implementation Strategy and Sequence

Strategy for Implementation of Improvements

We have identified 17 opportunities to improve process performance based on what we learned from the current state VSM. All 17 had benefit, most of them significant benefit. Although some were difficult, none was impossible; almost all offer enough potential that they are worth pursuing. In addition, Lean teaches us not to tolerate any waste, but to be diligent in weeding it out, thus providing more motivation to pursue even the more difficult opportunities.

Therefore, the question now is how to go about capturing the opportunities. There is often the tendency to go after the ones offering the most benefit first, especially those with high feasibility of success. That must be balanced against the reality that some improvements will be much more successful if delayed until some other, perhaps less individually beneficial, improvements have been completed. Some of the lower value improvements set the stage for the higher value improvements by creating a foundation on which they can be more easily implemented. For example, mothballing one baler (Opportunity 11) is one of the more economically beneficial opportunities we found, but it won't be practical until the reliability of the other balers has been improved (Opportunity 8) and their changeover times reduced (Opportunity 9).

Therefore, an implementation strategy is needed, one that will drive implementation in an appropriate sequence but still capture the very high value improvements relatively quickly. There is also a need to stage the improvements on a schedule that doesn't overwhelm the people who are required for implementation and cause them to lose focus; they can't be asked to do too many things at one time. We also want to avoid implementing an improvement when the work will have to be repeated after another improvement is completed. For example, most of the product wheel work could be done early, but would have to be redone after virtual cells are in place because that may change the product lineup on each piece of equipment.

The basic principles for sequencing improvements are as follows:

■ There is usually a logical sequence in which completed improvements set the stage for other improvements, where having some of the improvements in place makes the further improvements easier to implement or provides a higher likelihood of their success. This logical sequence should have the highest priority in determining the overall order of implementation.

■ High benefit opportunities should be done as soon as is practical, but not at the expense of following a logical path.

■ Things should be implemented in a sequence that eliminates re-work. For example, product wheels should be designed after cells are in place because the cellular design process will likely change the allocation of products to specific pieces of equipment and thus the wheel designs.

■ Implementations should be phased out as needed to avoid overloading the implementers.

A multi-generational future state plan is generally developed whenever there are a large number of opportunities to be pursued, like the 17 with Riverside Fibers.

Riverside Fiber Plant Future States

Figure 14.1 shows the complete list of opportunities found on the Riverside Fiber current state VSM, sorted first by benefit and then by feasibility. The benefit ranking ranges from 1 to 5, where 5 represents the highest benefit, 3 a moderate benefit, and 1 a very low benefit. In scoring each benefit, non-economic factors were considered, such as flow improvements and lead time reductions. That's why Opportunity 14, virtual cellular flow, got a benefit score of 4 even though it has no calculable economic benefit. Most opportunities showed some degree of economic benefit, and for them the ongoing benefit was given much more weight than the one-time benefit, for obvious reasons.

Feasibility was also scored on a 1 to 5 scale, where 5 represents a very easy implementation and 1 a very difficult implementation or one with low likelihood of success.

The timing is also shown; "now" means that the opportunity could be approached now, but could wait if there is a good reason to delay.

Figure 14.2 shows the benefit and feasibility scoring as an X–Y matrix. This is useful in communicating the results of this analysis with others; it makes it obvious that Opportunity 13 (mothball one polymer reactor) has very high benefit with very little difficulty foreseen in accomplishing it. In contrast, Opportunity 10

NUMBER	OPPORTUNITY	ONE-TIME BENEFIT (K$)	ON-GOING BENEFIT (K$)	BENEFIT SCORE	FEASIBILITY	TIMING
13	Mothball one polymer reactor	$0	$1,200	5	5	Now
7	Spinning changeover losses are high; spinning utilization is high	$0	$1,600	5	3	Now
12	Hot roll draw-anneal - reliability and yield wastes	$0	$435	4	4	Now
14	Implement Virtual Cellular Flow			4	4	After 8, 9, 11,12,13
11	Mothball one Baler	$0	$600	4	2	After 8 & 9
8	Baler reliability is poor			4	2	Now
9	Changeover improvement - Balers			4	2	Now
6	Bale finished product inventory is too high	$525	$140	3	5	Now
2	Flake inventory is high	$365	$998	3	5	After Cells
5	Right size the cutter box inventory	($12)	($4)	3	5	Now
15	Filamen tubs FIFOs after implementation of virtual cells	$425	$114	3	4	After Cells
1	Raw material inventory in rail cars is too high	$240	$65	2	5	Now
3	Filament inventory is high	$200	$56	2	5	Now
4	Rope finished product inventory in gaylords is high	$60	$16	2	5	Now
17	Drop the 60 very low demand SKUs	$60	$16	2	4	Now
10	Reduce the baler campaign cycle	$150	$40	2	2	After Cells
16	Implement pull replenishment across the value stream	The benefits were accounted for in opportunities 1–6				
		$2,000	$4,400			

	5-High	5-Easy
	3-Moderate	3-Moderate
	1-Low	1-Difficult

Figure 14.1 Opportunities sorted by benefit and feasibility.

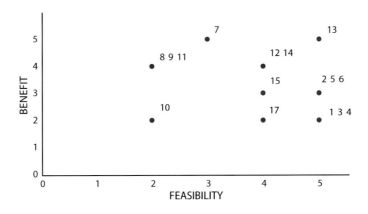

Figure 14.2   Benefit–feasibility matrix.

(reduce the baler campaign cycle) has relatively low benefit and low feasibility. That contributed to the decision to drop this opportunity, coupled with the fact that it and Opportunity 11 can't both be done.

Because of the number of improvements to be implemented, we have decided to stage them in three phases. Therefore, we have defined a future state 1, incorporating the first generation of improvements, and a future state 2 and 3. The far right column of Figure 14.3 lists the future state generation for each opportunity, and the chart has been re-sorted by generation.

Future State Generation 1

Future state 1 will include the highest benefit opportunities: mothballing one polymer reactor, getting rid of the hot roll draw–anneal machine, and reducing spinning changeover losses. These do not depend on any of the other opportunities, so there is no reason not to start immediately. The projects will be driven by Operations, with support from Technology in the process development work required to run hot roll products on the steam anneal machines and in seeking ways to further reduce polymer flow during spinning changeovers.

The remainder of the Generation 1 implementations is the inventory reductions achieved by bringing inventories into balance with that needed to support current performance. Inventory reductions relying on other successful implementations are in later generations. These improvements will be owned by the Production Planning and Scheduling function.

Figure 14.4 illustrates how the opportunities making up future state Generation 1 line up on the Benefit–Feasibility matrix.

Future State Generation 2

Future state 2 will focus on the balers. The primary goal is to eliminate one baler (opportunity 11); but for that to be possible, the performance of the other balers

NUMBER	OPPORTUNITY	ONE TIME BENEFIT (K$)	ON-GOING BENEFIT (K$)	BENEFIT SCORE	FEASIBILITY	TIMING	FUTURE STATE GENERATION
13	Mothball on polymer reactor	$0	$1,200	5	5	Now	1
7	Spinning changeover losses are high; spinning utilization is high	$0	$1,600	5	3	Now	1
12	Hot roll draw-anneal — reliability and yield wastes	$0	$435	4	4	Now	1
6	Bale finished product inventory is too high	$525	$140	3	5	Now	1
5	Right size the cutter box inventory	($12)	($4)	3	5	Now	1
1	Raw material inventory in rail cars is too high	$240	$65	2	5	Now	1
3	Filament inventory is high	$200	$56	2	5	Now	1
8	Baler reliability is poor			4	2	Now	2
9	Changeover improvement - Balers			4	2	Now	2
11	Mothball one Baler	$0	$600	4	2	After 8 & 9	2
17	Drop the 60 very low demand SKUs	$60	$16	2	4	Now	2
14	Implement Virtual Cellular Flow			4	4	After 8 & 9	3
2	Flake inventory is high	$365	$98	3	5	After Cells	3
15	Filament tub FIFOs after implementation of virtual cells	$425	$114	3	4	After Cells	3
4	Rope finished product inventory in gaylords is high	$60	$16	2	5	After Cells	3
10	Reduce the baler campaign cycle	$150	$40	2	2	After Cells	Drop

5-High	5-Easy
3-Moderate	3-Moderate
1-Low	1-Difficult

Figure 14.3 Opportunities sorted by future state generation.

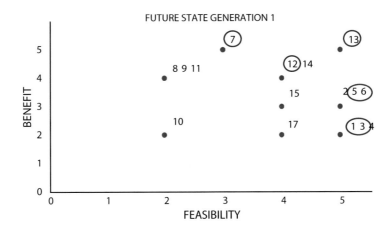

Figure 14.4 Future state Generation 1 benefit–feasibility matrix.

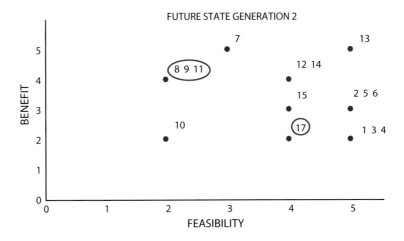

Figure 14.5 Future state Generation 2 benefit–feasibility matrix.

must improve. Therefore, this generation includes the baler reliability improvement effort and the changeover reduction program. If those meet their goals, baler OEE will improve to 66% and utilization will drop to 64%, making the mothballing of one baler practical.

The Operations function will own all of these programs, so this work must wait until the Generation 1 improvements requiring Operations support (13, 7, and 12) have been successfully completed.

Generation 2 also includes getting agreement from marketing and business management to drop the 60 obsolete products for the portfolio.

Figure 14.5 illustrates how the opportunities making up future state Generation 2 line up on the Benefit–Feasibility matrix.

Future State Generation 3

Implementing cellular manufacturing and the opportunities benefitting from the cellular flow patterns and product allocation are the focus of Generation 3.

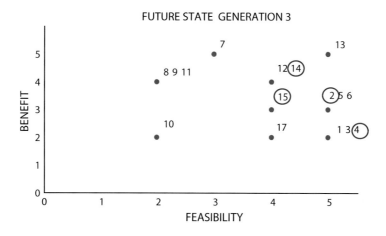

Figure 14.6   Future state Generation 3 benefit–feasibility matrix.

This includes Opportunity 2, implementing product wheels on the polymer reactors, and Opportunity 4, where all rope products will flow through one steam annealer. The cellular flow is also the enabler for the FIFOs between spinning and annealing.

Figure 14.6 illustrates how the opportunities making up future state Generation 3 line up on the Benefit–Feasibility matrix.

The cellular flow includes work done by Operations to determine the best flow paths, and by Technical to decide on the most appropriate product allocation for each cell. Production Planning and Scheduling will determine the new inventory targets for the polymer reactors and the rope steam annealer.

The multi-generational plan laid out above is logical, captures the greater benefits quickly, and phases out the workload for the affected work groups. Thus, it satisfies the principles listed earlier. There are certainly other sequences that will meet the criteria, so this is just one of perhaps several suitable implementation paths. For example, the implementations could have been spread out into more phases, to smooth out the work for Operations and Technical. Representatives of those functions should be heavily involved in determining the future state implementation sequence and phasing because they best understand their workload and limitations. People from other affected functions should also have a voice in this planning.

Summary

All of the effort that has gone into the development of the VSM will have been wasted unless actions are taken to resolve the problems seen and reduce the wastes highlighted on the map. Implementation planning is a critical step in the process, and has very significant influence on how successful the improvements will be, and ultimately how valuable the mapping work has been.

The tendency to go after the highest benefit opportunities first must be resisted, unless they occur early in a well-developed, logical plan. Implementation activities must be scheduled in a way that early successes build a foundation for the later improvements, and later improvements will not cause earlier improvement tasks to have to be repeated.

Now that a thorough, logical implementation plan has been developed, the next step is to generate future state VSMs to illustrate how the process will behave as each generation of improvements is completed. That will provide a clear picture of why the work is being done and of its value in enhanced process performance, and thus provide strong motivation to move ahead with the improvement tasks.

Chapter 15

Future State Value Stream Maps

Why a Future State VSM?

In the last chapter, we built a three-generation plan to implement the key improvements following a logical path, one that will get the highest benefit opportunities in place quickly and in a way that won't overburden the people required to do the work. Future state VSMs are an excellent way to illustrate the changes, how they will improve process flow, and how process performance will benefit. These maps create a vision of what a much more waste-free process will look like and how it will behave. The vision depicted on these maps provides a reminder of the specific performance goals we are seeking and thus a strong motivation to work diligently to make them a reality.

The maps also provide a framework to discuss and evaluate any additional changes being considered that are not based on the VSM analysis but were brought forward independently. Examples could include adding a new product family to the lineup, developing improvements to the steam anneal process, or cutting and baling product for customers who currently buy our rope to cut themselves and now want to outsource that operation. The future sate map is a baseline that can be used to calculate the effect any of these changes have on Takt, utilization, number of changeovers and thus OEE, and inventory levels.

The value of a multi-generational plan, with the various improvements grouped into a number of discrete phases, is that it keeps the implementation teams focused on a limited number of things at any one time, and gets them accomplished in a coordinated sequence of interrelated improvements. The alternative, a list of prioritized projects to be done whenever resources become available, can tend to rob the overall improvement effort of a systems perspective and make it less likely to understand the synergy of combinations of improvements.

An important reality: things beyond our direct control may (will!) change during the improvement process. There are external forces apart from our improvement efforts causing change. It is important to keep the maps up to date, both with the opportunities as they are implemented to reflect the improvements in process performance, and to capture other changes caused by events beyond our efforts.

The future state map for the generation currently being implemented should be updated with actual performance data as improvements are successfully completed. The next generation future state map should also be updated if any differences from the current generation targets affect that next generation.

The excitement and energy to move ahead with each successive future state should not bypass good change management and standard work principles. As changes are completed, they must be documented and communicated. New procedures must be written where appropriate, with training of all affected work groups.

Future State 1 Map

Figure 15.1 shows the future state VSM for Generation 1, with all changes highlighted in color. Some of the more significant features of this generation are:

- The role of the Production Supervisor has been redefined. He or she is to become much less of a supervisor and more of a coach to the people reporting to him or her. He or she will focus less on managing the equipment and much more on managing flow of material through the process. Visual management systems will be put in place to facilitate this transformation. This person will spend more time out on the floor reviewing the flow information on the visual display boards and assisting the operators in getting problems resolved. In recognition of the redirected responsibilities, the title will change from Production Supervisor to Flow Manager.
- The raw material ordering process will change. Suppliers will continue to receive a monthly forecast of material needs, but only for their information and for their planning processes. Material releases will now be generated by current rail car inventory status, in accordance with pull principles. The raw material suppliers are sent a signal when the adipic or diamine level drops below an order point. Rail car inventory has been cut in half because of this improvement.
- One of the polymer reactors has been taken out of service, and the campaign cycle of the remaining two has been cut to two days. This would allow a significant reduction in flake inventory in the silos, but that has been postponed until later because the people setting and managing inventory targets are busy with some of the other improvements.
- The hot roll draw–anneal machine has been dismantled and removed. Its products are now run on the steam anneal machines. This has improved the combined average yield of the draw–anneal step, so we now need slightly less material coming in to draw anneal because we waste less of it. Thus, the Takt of the filament coming from spinning has been reduced slightly.

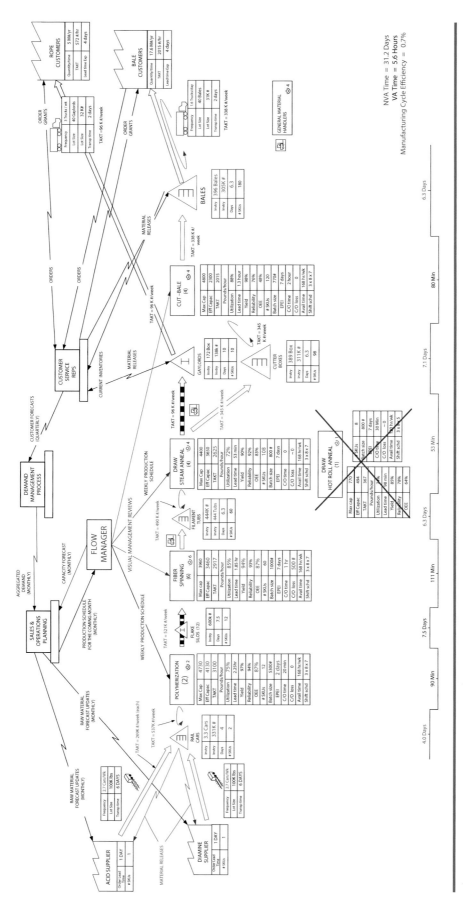

Figure 15.1 Future state Generation 1.

- Spinning machine changeover times have been cut in half, dramatically increasing spinning yield. Consequently, Takt of flake required by spinning has been reduced.
- In addition to the rail car inventory, the filament tub inventory, cutter box inventory, and bale inventory are all now managed as supermarkets in accordance with pull principles. Filament and bale inventories have been substantially reduced. As discussed earlier, cutter box inventory had been marginal in the past, so it has been increased slightly.
- To avoid overloading the Production Planning section, we have not implemented pull all the way through the process yet. Flake production and rope production for Gaylords will continue to be based on forecasts for now; pull will be implemented on those products in future state 3.

Some numbers in the data boxes may be different from those that were listed in Chapter 13 when each opportunity was being scoped, because of the synergistic benefit of making several improvements simultaneously. For example, when the polymer reactor opportunity (13) was scoped, it was predicted that with only two reactors online, the utilization would be 81%. We see now that it will be 75% because the Takt has been lowered by the spinning changeover improvement. When calculating the benefits of potential opportunities, we must treat each one separately and not assume that any others will be implemented. Therefore, it is not unusual for the numbers to improve when opportunities are implemented together.

The combined effect of all of the future state 1 improvements is that overall plant yield has increased from 75% to 81%, and lead time has decreased from 46 days to 31 days. Inventory worth $950,000 has been taken out of the system, and annual operating cost has been reduced by $3.5 million (Figure 14.3).

Future State 2 Map

Figure 15.2 shows the future state VSM for Generation 2, with the changes implemented in this phase highlighted in color. This generation is focused on the balers, including the reliability improvements, the changeover time reduction, and taking one baler out of service. With only three balers in use, the utilization rises to 91%. That is on the high side of the comfort level, but the benefits are worth trying to make it work. It should be emphasized that mothballing a baler is practical only if the reliability improvement and changeover reduction efforts meet their targets. Therefore, there is a possibility that this won't be feasible. This is typical of some future state goals. We have set a vision of a much more waste-free future and will try diligently to reach it, but we may not be successful in achieving some improvement targets. The low feasibility of any of these improvements should not deter us from trying.

If the reliability and changeover improvements don't achieve their targets, then four balers will still be required. In that case, Opportunity 10, reducing the

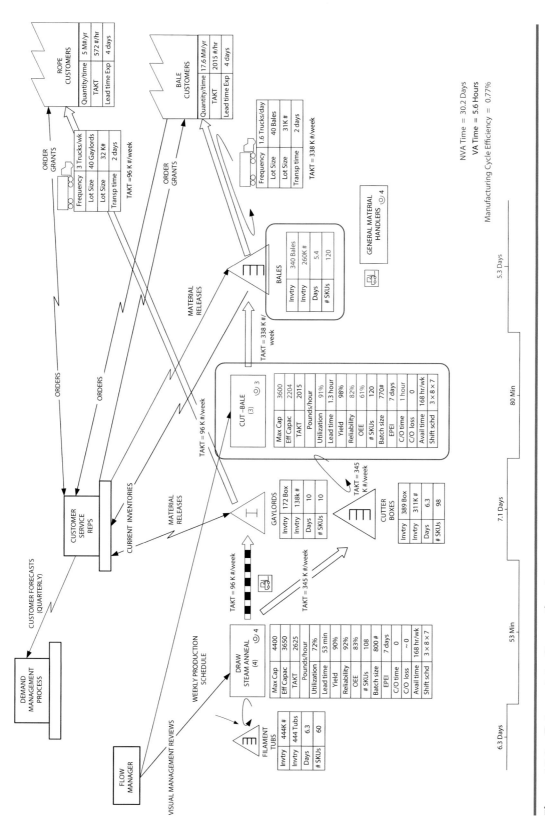

Figure 15.2 Future state Generation 2.

baler campaign cycle, should be reopened. In addition, the cellular flow patterns planned for future state 3 must be changed because four balers will be shared among the three bale producing cells.

Future state 2 also includes the elimination of the 60 slow moving or obsolete SKUs, so we can see that the bale finished product inventory has been reduced from 305,000 lb to 260,000 lb, taking out one day of non-value-add time.

The primary financial benefit of future state 2 is the $600,000 reduction in operating cost.

As future state 1 is implemented, some things will work out better than expected, and some not as well. Therefore, as future state 1 is being completed, the future state 2 map should be revised to reflect those differences. The future state maps should be considered living documents, and should be kept up to date as improvements are accomplished and as factors change.

Future State 3 Map

Putting the cellular flow plan in place is the most significant change in future state 3. This includes defining the new flow paths and visually identifying them so that there won't be any confusion about the new material movement routes. This will streamline flow and make it far easier to manage. Equally important is the grouping of products into families with similar processing characteristics and dedicating each family to a specific cell. Because of the similarities in processing requirements within a cell, changeover time and losses are generally substantially reduced. Figure 15.3 illustrates the cellular concept and flow paths. Because not all of the equipment is in multiples of four, some equipment must be shared between cells. Polymer reactor 1 is shared between cells 1 and 2, as is spinning machine 2. Even though some compromises have to be made, this is still a dramatic reduction in flow complexity; the total number of possible flow paths has gone from 192 down to 8.

Future state 3 also includes implementing product wheels on the polymer reactors, the spinning machines, and the balers. We waited until Generation 3 to take this step because product wheel design is based on the specific products run on each piece of equipment, and with the cellular product allocation, we now know what they will be. If product wheels had been designed earlier, they would have had to be redesigned now.

With this generation, we are completing the pull flow through the entire process by managing the flake silo inventory and the Gaylord inventory as supermarkets, and the filament inventory as FIFOs.

With those changes in place, the overall plant lead time has been reduced to 18 days, compared to a current state lead time of 46 days—a 60% reduction! This makes the Riverside Plant a much more responsive link in the Carolina Fiber Corp. supply chain.

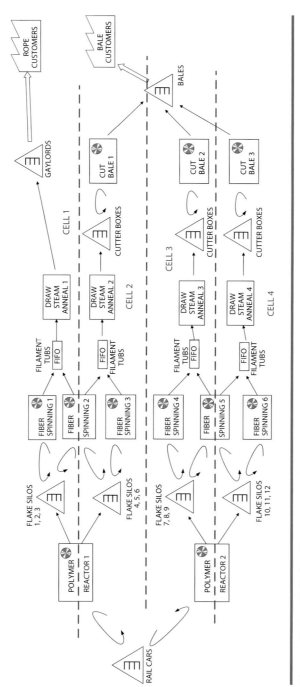

Figure 15.3   Cellular flow concept.

The reasonably predictable financial benefits of Generation 3 include $850,000 reduction in working capital resulting from the inventory reductions, and an ongoing reduction of $228,000 in operating costs. There will also very likely be a substantial benefit coming from the cellular flow product allocation, but we won't be able to estimate that until the allocation is done. For example, the product allocation on any one of the spinning machines may result in a simpler combination of changeovers, thus lowering the changeover losses. The product wheel design may find a sequence to schedule the spinning products, which will also lower the changeover losses. These further improvements can't be quantified at this time, and so are not reflected on the future state VSM, but are a real likelihood.

Figure 15.4 illustrates future state Generation 3, with the improvements new to this generation highlighted in color.

Some observations from this map and the cellular concept are:

■ Cell 1 will have slightly lower Takt than the lower cells because the Takt of rope going into Gaylords is less than 1/3 of the bale Takt. This is not an important distinction for a future state VSM because the product family allocation is not designed yet, and when that is done, it will likely create larger (but manageable) Takt imbalances than that.

■ Because we are now fully on pull and all scheduling is done by product wheels, there is no need to send the weekly schedule to the plant floor; the weekly schedule is now set by the product wheel design and current inventory status. If there are significant changes in demand for any product, the Production Planning function will update the cycle stock values used to set the inventory targets on which the pull signals are based.

■ Filament tub inventory managed as FIFOs is shown as a half day, 36,000 lb. In reality, it will typically be somewhat smaller.

Summary

Future state maps are a very important outcome of the analysis of the current state VSM, and are a critical contributor to the motivation to move forward with improvement tasks. They provide a visual roadmap of the improvements to be achieved in logical sequential groups. The future state map for whatever generation is currently being implemented should be kept current as improvements are completed, and prominently displayed to keep everyone informed of progress and status.

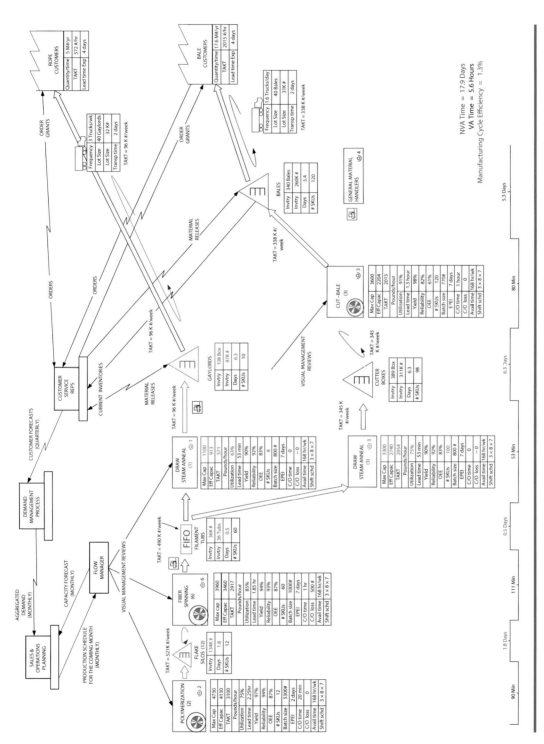

Figure 15.4 Future state Generation 3.

Chapter 16

Supply Chain Mapping

Why a Supply Chain Map Is Important

Understanding flow and waste is just as important for your entire supply chain as it is for your manufacturing operations, and often more so. Supply chain waste can often be much higher than the wastes found in the individual operations, and can have more significant implications for system flexibility, customer service, and overall business performance.

A Supply Chain Map (SCM) is an excellent way to see the waste and envision what can be done to reduce it. It uses a mapping format very similar to a VSM, but will be at a somewhat higher level, and show each entire manufacturing operation as a single process box. There will be less emphasis on equipment level metrics, and more on plant-wide metrics. For example, Takt, capacity, and yield will be generalized for the entire plant, rather than being focused on specific pieces of equipment or production lines. In addition, the lead time across the entire plant will be shown.

In addition to each manufacturing facility, the map will include each warehouse and distribution center (DC), all other inventories external to the plants and warehouses including inventory in transit, all major sources of material supply, and all major customer groups. Information flow will be shown; it plays a critical role in understanding how the supply chain is being managed and is just as important to an SCM as it is to a VSM.

Because supply chains have traditionally gotten less focus than manufacturing operations, there is generally much more waste and much more opportunity to be captured.

Supply chains have been traditionally managed as individual components (production facilities, warehouses, distribution centers, etc.) rather than as integrated flow chains, so there are usually a number of previously unrecognized flow issues that an SCM can bring to light. For example, in many enterprises, logistics have been optimized on cost rather than on lead time. Selecting modes of transportation and providers based on lowest cost makes economic sense on the surface, but probing deeper often reveals a different view. Frequently, the lowest cost transportation options do not guarantee the shortest travel distance or the shortest lead time. Hence inventory and inventory carrying costs may increase beyond the savings in transportation. Worse, longer lead times make the system less flexible and less able to react to changes in customer needs. An SCM puts these trade-offs in clear focus.

Supply Chain Wastes

Several of the seven wastes that Ohno articulated for manufacturing can also be found beyond the manufacturing walls, across the entire supply chain.

1. **Inventory**—This is one of the biggest wastes we see across supply chains. Supply chain inventory waste includes inventory in warehouses and DCs, and so-called "pipeline inventory," inventory in trucks, railcars, and on ships, whether it is moving or parked in a rail yard.

   All inventory is waste, but some is necessary waste, required to ensure smooth flow of our products to our customers. The SCM will help you see the total inventory, with enough operating data to estimate if it exceeds the level necessary for smooth flow. It will also help you understand how performance must be improved to reduce the necessary inventory.

2. **Transportation**—All transportation is waste, but some transportation is inevitable because we can't locate our raw material suppliers and manufacturing facilities adjacent to our customer's location. Given that some transportation is required, there are three additional causes of this waste: (1) inappropriate location of warehouses and DCs, (2) longer travel distances than required because we have chosen an inefficient transportation method or route, or (3) moving material back and forth between warehouses because poor planning has located specific SKUs in the wrong location.

3. **Defects**—The primary supply chain-wide defects are accounting errors, which lead to inventory inaccuracy; this may be caused by problems related to the Warehouse Management Systems (WMS) for the individual warehouses, or it may have system-wide causes. At any rate, it will have system-wide ramifications. Overstatement of the inventory of a material can cause stockouts, missed shipments, and poor customer service. Inventory errors that understate the inventory will cause excess inventory to be maintained, incurring extra cost.

4. **Extra processing**—The most likely form of this type of waste is extra information processing at the supply chain management level, which may be seen from the SCM information flow. It may cause delays in information getting to the right users in a timely manner and thus delay the material flow.

Another example of extra processing waste would be extra handling in warehouses, loading and unloading ships, railcars, trucks, etc., but these are local wastes and would not be seen on an SCM, and would require a VSM of each warehouse, DC, and shipping and receiving station to become apparent.

5. **Overproduction**—Ideally, the entire supply chain should have integrated flow control based on pull principles. If not, it can cause any individual manufacturing facility within the chain to overproduce.

6. **Movement**—This refers to operator movement, so there is little of this at the supply chain level.

7. **Waiting**—This refers to operators having to wait for input materials or for equipment to become available; as above, this is negligible at the supply chain level.

Effects of Wastes at the Supply Chain Level

The primary ramifications of poor supply chain performance caused by these wastes are:

Poor customer service levels: If supply chain inventories are not well managed, you can be in the seemingly paradoxical situation of having too much inventory but not being able to meet customer needs because inventory is in the wrong product mix or at the wrong location. In addition, if there are flaws in the inventory recording and accounting processes, that can lead to inaccurate inventory information, which may cause you to think you have enough inventory in any particular SKU, while in reality it is not there.

Excess inventory and inventory carrying cost: As noted previously, if inventories are not well managed, you may have far more inventory than you need for smooth, continuous flow of your products. And that inventory costs you, both in the carrying cost of the capital tied up and in the cost of the facilities to store it.

Long lead times and lack of flexibility: If a supply chain is not well designed, if the trade-offs between logistics or manufacturing cost and lead time are not analyzed carefully, you may end up with a lot of intercontinental transportation and very long lead times. Long lead times rob you of flexibility and inhibit your ability to react quickly to changes in customer demand.

Excessive transportation costs: If off-shoring decisions are not made with a complete view of all costs, you may end up with inexpensive production costs but high transportation costs, long lead times, and high "pipeline" inventory.

Any of these can be extremely detrimental to an enterprise, and a combination of them could be catastrophic, so they must be minimized. An SCM is an excel-

WASTE	DESCRIPTION	EFFECT
Inventory	Excessive inventories due to lack of an integrated supply chain pull system, and to inefficient modes of transportation	High Inventory costs
Transportation	Excessive distances due to non-optimal routing; high cost per mile due to poor selection of mode of transportation	High transportation cost
	Long transport times due to non-optimal routing	Long lead times and lack of flexibility
Defects	Inventory errors; inventory inaccuracy	Missed shipments and poor customer service, or extra inventory
Extra Processing	Unneeded information handling and processing in the supply chain management operations	Delays in information transmission, causing delays in material flows
Overproduction	Excessive production within manufacturing operations due to lack of an integrated supply chain pull system	Extra inventory
Movement	Not at the supply chain level	None
Waiting	Not at the supply chain level	None

Figure 16.1 Supply chain wastes.

lent way to understand the wastes and their root causes so that these problems can be minimized.

Ohno's seven wastes, how they manifest themselves in a supply chain, and the negative effects of these wastes are summarized in Figure 16.1.

Supply Chain Map Components

An SCM has the same major components as a VSM: material flow with data boxes, information flow, and a timeline. The kind of data in the data boxes is similar, but at a higher level, and with less detail.

Each production facility should be shown as a single process box. The accompanying data box should include parameters that describe the performance of the facility as a whole: the total plant Takt, effective capacity, and utilization. The overall lead time through the operation and the number of SKUs produced are generally included. OEE may also be shown if it adds useful information, but the OEE value for the entire facility can be very misleading if there are lines in parallel within the facility, or if there are internal buffers de-coupling any of the process steps. Keep in mind that the purpose of the data boxes in this case is not to diagnose operational difficulties within any plant, but to understand their effect on supply chain flow and performance.

Figure 16.2 shows how the Riverside Fiber Plant could be shown on an SCM. Most parameters in the data box summarize the entire plant performance. However, Effective Capacity, Takt, and Utilization reflect the bottleneck or near-bottleneck operation, which in this case is Spinning because that is what limits

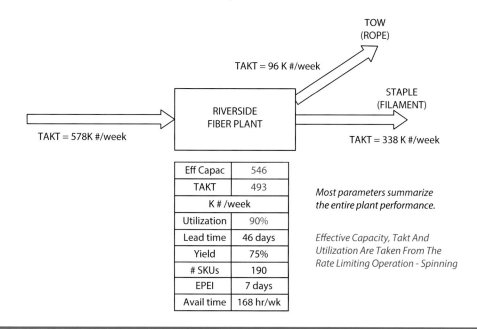

Figure 16.2 The Riverside Plant as it would appear on an SCM.

flow through the entire facility. OEE is not listed; an overall plant-wide OEE figure would be relatively meaningless because of the internal inventories, which de-couple various equipment downtimes. OEE is more important on a VSM, when analyzing the internal plant operations for performance issues; the data shown on the SCM are important in defining how well the Riverside Plant performs its role as a link in the supply chain.

One of the ground rules for drawing a VSM, that you keep it at a high enough level of detail that the entire operation can be shown in perhaps 6 to 12 process boxes, is difficult to follow with an SCM. A supply chain typically has enough major nodes that the map must necessarily occupy a lot of real estate. Every production site, every warehouse or DC, all major supplier groups, and all significant customer groups should be shown in order to provide a visual description of flow and its inhibitors.

If there are many parallel paths in the supply chain, often the case, showing everything on the timeline can get messy. It is not necessary to represent each parallel activity on the timeline. It is generally important to show the timing for the flow of the most significant materials, and to show the longest path through the supply chain. This is another case where your judgment is critical; the timeline should include whatever is required to give the best illustration of how responsive the supply chain is, and where the major holdups occur.

Future State Supply Chain Map

An SCM is virtually useless unless it becomes a blueprint for action, unless you plan to move forward to rectify the problems that the map illustrated. Toward that end, a future state SCM serves the same purpose and provides the same value that a future state VSM does. It provides a vision of what a far less wasteful supply chain would look like and the performance improvements that would enable. Documenting all of the operational benefits to be achieved by reaching the future state in one place provides tremendous motivation to create and follow an action plan.

Supply Chain Map Example

A very simplified portion of a global SCM is shown in Figures 16.3a, 16.3b, and 16.3c. A lot of the detail has been removed, and the map has been separated into three sections to enable printing at a reasonable size. The plant names and some of the process material designations have been changed in accordance with non-disclosure agreements.

The end products of this supply chain are various grades of plastic materials that serve a wide variety of end uses. The processing is based on fluorine chemistry, and begins with an ore containing fluorine being mined in China, Africa, and South America. The ore is shipped to North America by barge and transferred to railcars to be transported to the first plant in the chain, where fluorine is extracted from the ore to make material R. Material R is transported by rail to Plant B to make material S. Material S is moved by rail to Plant C where the polymerization process occurs, producing the plastic materials sold to external customers. Plant C also receives material V, also required for the polymerization reactions, from two other plants.

One of the most obvious features of this SCM is the large number of triangles; inventory is everywhere! Not only is there the required pipeline inventory on barges and in moving railcars, but also there is an equal or greater amount in parked railcars. In some cases, these parked railcars contain inventory being deliberately held at that node in the supply chain, but in other cases, it represents inefficiencies in railcar management.

What we learned as we were creating the map by interviewing the various people responsible for managing this supply chain was that every inventory pile was well understood by someone, but it was a different someone for almost every inventory, and they didn't talk to each other very often. Therefore, no one had a view of the big picture; no one understood how much inventory was tied up in the system. The SCM was a real eye opener, as it was the first time anyone had seen the global flow on a single sheet of paper. The total lead time through the entire supply chain was 154 days, about 5 months! Of that, only about 30 hours were value adding, for a Supply Chain Efficiency of less than 1% (Supply Chain Efficiency = Value Add time/Total Time).

As we drilled down into the details, the following things became clear:

■ Because each inventory location was being managed by a different person, with little visibility across the chain, there was extreme duplication

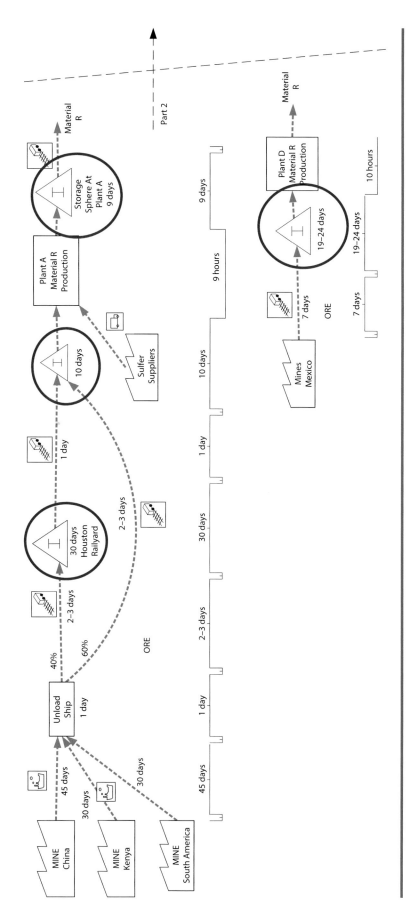

Figure 16.3a Supply chain example—part 1.

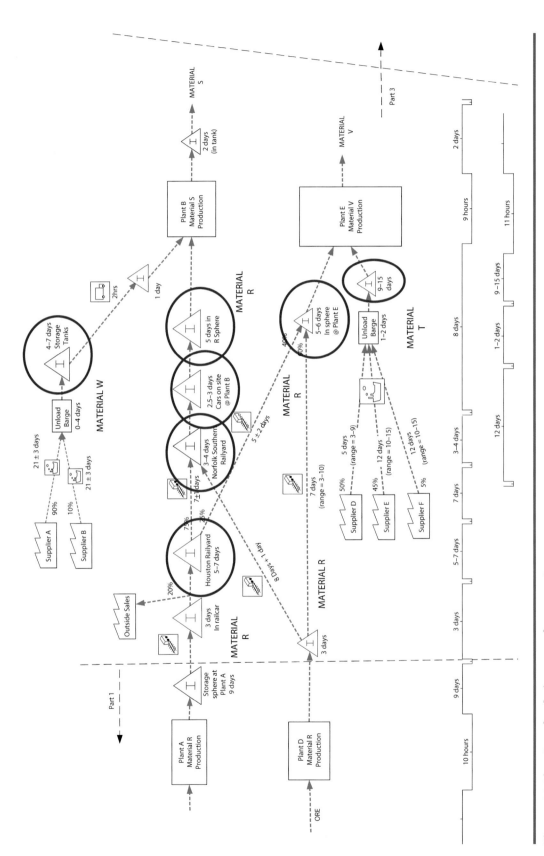

Figure 16.3b Supply chain example—part 2.

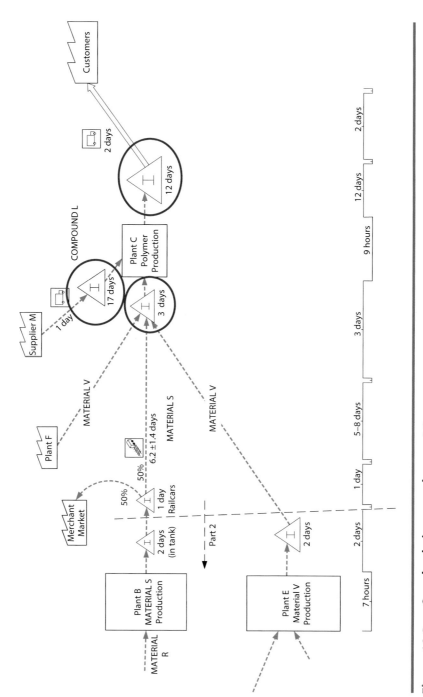

Figure 16.3c Supply chain example—part 3.

of inventory, both in cycle stock and safety stock. The amount of inventory stored at each plant was far more than required to satisfy production cycles.

■ Because rail contracts were negotiated on a lowest cost basis rather than on a performance basis, railcars would sit in rail yards waiting switching for quite a number of days. The lead time wasted in this manner averaged 40 days, about 25% of the total lead time.

■ The monthly Sales and Operations Planning (S&OP) process had been viewed as running well and effectively managing production. However, this analysis revealed that while it worked reasonably well at the monthly level, it tended to be ineffective during shorter time intervals. Each of the interconnected plants that supplied each other felt that they had the latitude to set their own daily schedules as long as they met the monthly requirement. This not only caused supply problems, but also it played havoc with the logistics planning process which was responsible for making sure that the railcar and tank truck fleets were in the required locations at any time.

To get a better understanding of what inventories were needed to manage the variabilities in the supply chain, a discrete event simulation model was built. Adjusting the inventories to the levels indicated by the computer model allowed a very significant reduction in working capital, almost $15 million, while at the same time improving supply chain delivery performance.

It is almost always the case that the first time an operations manager sees his or her entire supply chain on a single piece of paper, he or she immediately sees problems that had been hidden by the complexity of the chain. Opportunities to improve performance, save cost, and improve responsiveness become apparent.

Summary

An SCM is every bit as critical to becoming a Lean enterprise as VSMs are. As you move beyond the internal details of individual supply chain components, the production facilities and warehouses, there are fewer categories of waste compared with Ohno's original seven. However, those that are there, principally inventory and transportation, can have a far greater impact on business performance than the wastes we see in manufacturing. Supply chain waste can add cost and lead time to your delivery processes and make you less responsive to your customers. And the number of inter-organizational handoffs that we see in a poorly managed supply chain leads to errors, missed shipments, and inappropriate schedule changes. Because of the traditional practice of managing supply chains as a collection of separate pieces rather than as an integrated whole, system-wide problems have frequently been invisible and unrecognized.

An SCM is a powerful way to see the supply chain as an interconnected operation, to highlight wastes and flow discontinuities, and to drive action toward improvement.

Chapter 17

VSM as a Way of Engaging Employees

Origin of the Problem

Throughout most of the twentieth century, the attitude about the workforce was that they were hired for their ability to do work: to operate machines, to monitor processes, and to perform maintenance tasks. Worker perspectives on how the work should be done, on how to correct defects in work processes, or how to improve the operation were not encouraged. This came to be known as the "check your brain at the door" culture.

This belief about the role of the workforce probably began with Frederick Winslow Taylor and the high degree of influence he had on manufacturers in the early part of the twentieth century. Taylor was one of the first to view work as a process, to measure and analyze work, and to promote work standards, and is widely regarded as the father of scientific management. One of his most insightful observations was that planning work and doing work were two completely separate activities. While this may seem obvious today, Taylor was the first to articulate it. The difficulty arose when Taylor concluded that these two activities should be done by different groups of people, that managers should design and plan the work and laborers should do the work. Taylor believed that the mental work and the physical work should be separated because those willing to do the physical work were incapable of planning and organizing the work.

> "In almost all the mechanic arts the science which underlies each act of each workman is so great and amounts to so much that the workman who is best suited to actually doing the work is incapable of fully understanding this science."

In fact, in remarks to a congressional committee, Taylor said:

> "I can say, without the slightest hesitation, that the science of handling pig-iron is so great that the man who is physically able to handle pig-iron and is sufficiently phlegmatic and stupid to choose this for his occupation is rarely able to comprehend the science of handling pig-iron."

And that such a man

> "… more nearly resembles in his mental make-up the ox …."

Although Taylor had some very inappropriate views on the capability of the labor force, his views on other topics regarding how work is done were considered groundbreaking and powerful. His principles of scientific management came to be known as Taylorism, and had a profound influence on other industrialists of his age. And so the paradigm on the separation of doing work and planning and improving work as set in place for decades to come.

A New Paradigm on the Role of Labor

Toyota realized the fallacy with this attitude, and promoted the idea that people are the most important asset in an operation, and need to be valued, developed, and nurtured. For example, Toyota's emphasis on continuous improvement was not only to improve process performance, but also to improve employee capability. Continuous improvement activities are an excellent way to teach analysis and problem solving skills. Toyota's success in manufacturing created many disciples, and so Toyota's attitudes about the value of the workforce began to spread, leading to the various employee engagement initiatives such as HPWS (High Performing Work Systems) and HPO (High Performing Organizations) that became so popular in the 1980s.

VSM development is one Lean activity that both promotes employee engagement and benefits from it. It taps into employee capability and enhances the development of cognitive insight and analysis and problem solving skills. Participation in VSM creation, analysis, opportunity scoping, implementation planning, and execution is an excellent way to develop these skills. This is far more than making people feel good about their contributions, or improving labor–management relations: there are real benefits from having operators, technicians, mechanics, and administrative people heavily involved in all the activities associated with VSMs:

- It reinforces that employee perspectives are valued, that operators and mechanics are valued for their intelligence as much as their muscles.
- Participating in map creation develops their observational and critical thinking skills.

- Participate in reading the map and looking for waste and for flow improvement possibilities develops their analytical skills.
- Participate in generating the future state maps develops their problem solving and project scoping skills.
- Exploring and scoping the improvement possibilities seen on the map helps to foster a Continuous Improvement mindset and culture.
- Participating taps into their process understanding and creates a more thorough, more accurate map.
- Participating in the opportunity prioritization and future state development builds a more thorough understanding of the changes required and a clearer picture of the difficulties that may be encountered. Line operators often have insight into what is practical to do on the plant floor, what could cause problems, and possible ways to mitigate them.
- Participate in the opportunity prioritization and future state development builds buy-in, support, and commitment.
- VSM creation and the use of the map as an ongoing analysis tool can be a very effective way to build stronger workforce–management interconnectedness around the manufacturing process. VSMs should be a strong component of worker engagement initiatives and of the visual management design.
- Finally, this sends a signal that the workplace culture is changing, that eliminating waste and simplifying work are the real goals, not cost reduction or headcount reduction. Cost reduction should be seen as an outcome of waste elimination, not a driver unto itself.

The Nature of Engagement

Employee engagement and empowerment can be summed up in five words that begin with the letter E (Figure 17.1):

- Engage—Make people feel like they are really a part of the organization, that they will share in any successes, that what is in the best interest of the business is in their personal best interest. It is very important that every

Figure 17.1 Employee engaging–empowering model.

employee clearly understand the corporate goals, the business goals for their plant or work unit, and how their individual actions contribute to successfully meeting those goals. People are highly motivated to do what is in their own self-interest, so helping them understand how business success will drive personal rewards can be a powerful force in engaging them in improving the operation. And the rewards should be in the form of recognition, increased job responsibility, and increased decision-making authority in addition to any monetary rewards. Insightful contributions to the creation of a VSM can be rewarded by giving more responsibility and more latitude in creating the future state, which is both engaging and enabling.

■ Enable—Give people the freedom to make changes to improve their work, as long as they operate within clearly articulated boundaries, and within the operation's Management of Change (MOC) process. Provide reasonable levels of funding for improvements, as long as they are justified by the benefits to be realized.

■ Enlighten and educate—Put processes in place to make sure that everyone is trained in their job tasks. I've seen operations where new employees were turned loose into the manufacturing operation without any specific training on the process or their specific tasks, and were expected to learn the job requirements on the fly. This can lead to false assumptions of what the job entails and how it should be done, and severely hampers best practices on Standard Work. Participation in VSM creation is no substitute for job training, but it will enhance good job training by giving participants a more thorough understanding of the end-to-end process and how their normal job tasks contribute to overall process performance. Employees should also be trained in quality tools and Lean tools like 5S, 5Whys, SMED, and pull replenishment concepts, so that they can envision improvement possibilities as they analyze the current state map.

■ Encourage—Encourage them to try different approaches, and not be afraid of failure. Improvements that could lead to a much better future state often entail a risk that they won't work out, and it's important to let future state VSM architects know that they have a safety net. If you are trying to improve, not everything will be successful, and that's acceptable. "If you're not failing sometimes, you're not trying hard enough." Celebrate well-intentioned failures as well as successes.

■ Empower—True empowerment requires providing the workforce with the right skills, the right level of encouragement, the right motivation, and the freedom to work within reasonable boundaries. With that, there's no limit to what people can achieve!

Paying specific attention to the Engage, Enlighten, Encourage, and Enable principles will result in higher quality, more thorough VSMs, and a better set of improvement possibilities to consider for the future state. The appropriate discipline around Standard Work and Management of Change practices is critical.

Success is highly dependent on the level of trust between managers and workers, so it is important that agreements and commitments be honored by both parties.

So, you can now add 5Es to your list of Lean processes based on the number 5—5S and 5 Whys.

Summary

The entire workforce, i.e., the operators, mechanics, laboratory technicians, quality control personnel, buyers, planners, and material handlers are all critical components of the success of the operation and therefore of business success. Failure to engage, to tap into their experience, insight, and creativity, is perhaps the biggest waste of all and is why many Lean practitioners have added this as the eighth waste alongside Ohno's original seven.

VSM creation, VSM analysis, and future state map development provide excellent platforms for greater engagement. It is a two-way street, a mutually beneficial practice, in that it promotes employee participation and engagement and builds a better result.

Chapter 18

A Roadmap for Continuous Improvement

To get full value from the work you have put into creating the VSM, it should be a living document and kept evergreen. The benefits achieved in reaching the defined generations of future state are only a part of the total benefit it can provide. It is highly unlikely that you will ever get all of the waste out of the process, so an up-to-date VSM will allow you to see what waste is still there, and why you don't yet have perfect uninterrupted flow exactly matching Takt. A VSM that accurately describes the current process parameters and performance may reveal waste and problems that had previously been masked by more significant problems but have now risen in visibility, and also indicate areas where recently completed improvements enable other improvements to be made. Thus, the VSM can provide the roadmap to guide you along the continuous improvement journey.

It is important that as each group of improvements is completed, as the next future state generation is reached, the map be updated to reflect those improvements and their effect on all map parameters. When a specific generation of the future state is reached, the process may not behave exactly like the future state map for that generation had predicted. Some of the scoped improvements will work out even better than projected, while some will not hit their targets completely, so the map should be updated to reflect the true process performance at that point in time.

It may also happen that parts of the process have improved by independent actions, not triggered by analysis of the VSM but done outside of and in parallel with that improvement framework. When each of these activities is completed, the results should be used to update the map. Examples could include:

- The technical group has reduced yield losses in one of the process steps. This should be shown on the VSM both in the yield parameter in the data box for that step and in the reduced Takt requirement for the upstream steps. As yield is a component of OEE, OEE will also improve, increasing Effective Capacity, resulting in a decrease in Utilization, which should all be updated.

◼ A cross-functional team including maintenance, mechanical engineering, and electrical engineering has improved the reliability of a key piece of equipment. This also will increase OEE and Effective Capacity and reduce Utilization.

◼ If a Kaizen event or a technical program can reduce changeover time, the new changeover time and the increase in OEE should be shown and, in addition, starbursts should be drawn around EPEI for that step and around the next inventory bucket to highlight that they should be examined to capitalize on the reduction in changeover time.

◼ If the ERP (Enterprise Resource Planning, e.g., SAP, Oracle) computer system that manages all data is upgraded or replaced with a newer system, the information flow portion of the VSM will very likely need changes to depict the new information processing. In some cases, business managers try to force fit their old processes into the new system, so not much will change in the VSM information flow. However, the more progressive companies recognize that the introduction of a new ERP system provides the opportunity to re-examine and improve all business processes, both to correct deficiencies in the old processes and to take advantage of the enhanced capability the new system offers. Therefore, in those situations, the VSM information flow will likely see significant changes.

Completely independent of any *process* performance improvements made as a direct result of VSM learnings or in parallel with them, characteristics of the *products* being made may change in a way that should be reflected on the VSM.

◼ It is generally the case that demand for some products will see slight increases while others see a drop. If these changes are small, the increases and decreases may more or less even out, so the overall Takt shown on the map stays relatively constant. However, if demand for a product or product family increases significantly or if a new product is added to the lineup, Takt values on the map should be updated. Also, if demand for a product or group of products drops dramatically, or if products are removed from the portfolio, any substantial reduction in Takt must be reflected on the map.

◼ Increases in Takt should trigger a re-examination of potential bottlenecks. Reductions in Takt should trigger an analysis of utilizations to see if some equipment could run a shorter shift schedule or be idled. Either should trigger a re-examination of the product allocation (Group Technology) portion of the cellular flow plan, if that strategy is being used here.

◼ The inventory levels shown on the map should be re-examined periodically to see if they are still appropriate for current process performance. If better demand planning results in reduced forecast error, less safety stock will be required. Similarly, if demand for a highly variable product stabilizes, it will have the same effect, in cases where demand history has a major influence on the forecast.

Another reason to keep the VSM up to date is its role in the visual management process; an up-to-date version of the map should be prominently displayed. There should be periodic reviews for all employees, to celebrate recently completed improvements, to highlight upcoming improvement efforts, and to solicit further ideas to reduce the wastes and flow barriers depicted on the map. As employees pass by the map in their normal path through the plant, they should be encouraged to stop, look, and think about it, and make recommendations of any ideas they have for resolving the starburst issues.

As new employees are hired, the current version of the VSM provides an excellent way to orient them on the manufacturing process, flow enablers and detractors, and wastes that should be addressed. More importantly, it provides you with a vehicle to explain to them how their tasks help to create value for the customer.

Summary

It is not realistic to expect that you will ever get all of the waste out of your processes; hence the need to frequently re-examine process behavior to see what additional improvements are possible. An up-to-date VSM provides an excellent tool for doing that, in a context that clarifies the effect of process improvements on process performance and flow, so it is critical that the VSM be updated as each improvement or set of improvements is completed.

Continuous improvement requires an up-to-date assessment of performance, detractors, and waste to know where to focus the next round of improvements, and by providing that the VSM lays out a roadmap for ongoing improvement.

## Chapter 19

Benefits of Developing a VSM

Operations That Have Benefitted from Using a VSM

A wide variety of processes, including those making automotive and house paints, food products, extruded polymers, paper and plastic sheet goods, industrial chemicals, engine oil additives, waxes and pastes, and laminated circuit board materials, have improved their operation using the VSM format. Here are a few examples to demonstrate the breadth of usage, and to illustrate some of the concepts covered throughout this book. They include situations where the benefits were the traditional ones, illustrating the primary wastes and flow problems in the manufacturing system, as well as applications where the benefits were unique to the particular situation in that operation.

Processing of Large Rolls in a Sheet Goods Plant

A VSM was the key to understanding how to resolve poor customer service performance for a sheet goods plant. The plant produced a paper-like polymeric product sold in large rolls of widths varying from 2 feet to 12 feet, and weighing 300 lb to 2000 lb. The irony of their problem was that they had 50 days of finished goods inventory, yet provided their customers only 84% on-time delivery.

The VSM reflecting the current state at the start of our program is shown in Figure 19.1. The incoming raw material, polymeric plastic pellets, is stored in railcars. It is melted and cast as mill rolls 10 to 12 feet wide and weighing approximately 2000 lb, comprising 50 SKUs. The mill rolls are stored in an AS/RS (Automatic Storage/Retrieval System, a high-rise warehouse). They are then taken out of storage and heat treated on one of three bonding machines and returned to the AS/RS. They are later removed and slit along the length of

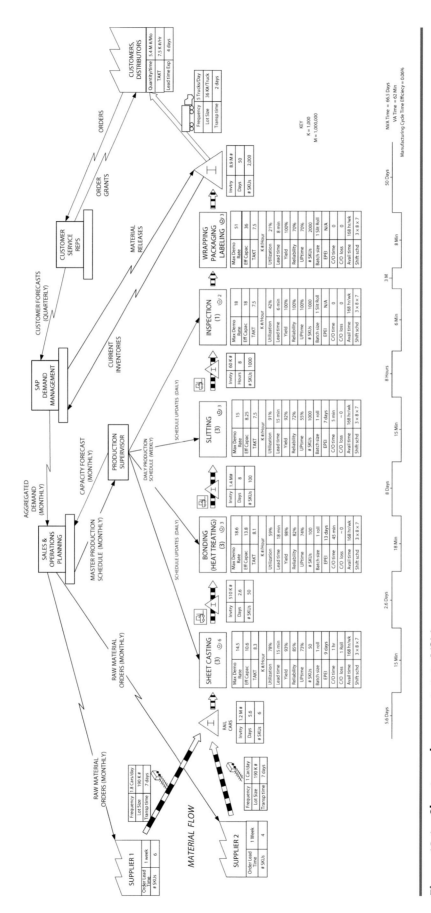

Figure 19.1 Sheet goods current state VSM.

the roll on one of three slitters, to create several narrower slit rolls from each mill roll. Three to six slit rolls result from each mill roll, depending on the slit width. There are approximately 1000 SKUs of slit rolls. Slit rolls are inspected, wrapped, and transported to an off-site finished goods warehouse.

An analysis of the make-up of the finished goods inventory and of the SKU ordering patterns done as part of the VSM data gathering revealed two things:

- Forecast accuracy was very poor, causing the mismatch between inventory and orders.
- The material in the warehouse was in the needed cast and bonded styles, but had been slit to unneeded widths.

The proposed solution was to stop slitting to the forecast to create finished goods inventory, and instead to maintain a larger pre-slit inventory and slit to fulfill specific orders. This would eliminate essentially all finished product inventory, replacing it with pre-slit inventory. The VSM showed that there was enough space in the AS/RS to hold the required amount of pre-slit inventory, which was less than the finished goods inventory. The forecast was far more accurate for bonded styles than for slit products, so much less safety stock was required.

The VSM also revealed, however, that the three slitters would not be able to keep up with the instantaneous level of demand required by a slit-to-order strategy. Slitter Uptime (their term for OEE) was only 55%; slitter downtime had previously been covered by the high level of finished product inventory, but the slitters would now have to be available to satisfy the short customer lead times required by any finish-to-order strategy. Discussions with operating personnel revealed that there had been a belief that there was excess slitter capacity, so slitter maintenance had not been a priority. Therefore, a key component of the future state plan was to improve slitter reliability. It was believed that with the appropriate level of maintenance, slitter reliability could be returned to the 78% to 80% levels that were experienced earlier in their lifespan.

Another possible slitter Uptime improvement would be to reduce the slitter changeover time. Repositioning the knife blades on the slitter to accommodate a different slit pattern took only 5 minutes, but with many of the 1000 SKUs produced every 7 days, these changeovers dropped Uptime by 16%. The team felt that better coordination of the slitting operators could reduce this to 4 minutes, giving a final Uptime of 62%. This wasn't included in the future state 1 plan, but was considered for later. The reliability improvement in future state 1 would drop Utilization to 83%, thus giving enough capacity to handle the peak volume of customer orders within any 24-hour period. (Customer lead time expectations were that orders would be received within 4 days; truck loading and shipment required 3 days, leaving one day to slit, inspect, and wrap rolls to meet the lead time target.)

Therefore, the benefit provided here was that the data gathering done to populate the VSM clarified the problem and its root cause, and the VSM highlighted the improvement in equipment performance required to execute a slit-to-order strategy.

Figure 19.2 shows the relevant portion of the future state VSM and the improvement that was expected when the reliability had been raised to the target level. The pre-slit inventory would have to be increased to have the required material available in the right types and in sufficient quantity to support the slit-to-order strategy.

The finished product inventory could go to just a few hours; the only material needed is that coming from packaging waiting to be loaded on a truck. However, keeping a few days of the very consistent, very high demand products in stock would relieve some of the pressure on slitting, inspection, and wrapping. If the demand on slitting temporarily got too high, it would likely include some of those SKUs, and they could be shipped from stock and replenished later.

The expected outcome was that the combination of pre-slit and finished product inventory would be reduced from 58 days to 22 days, a 62%-decrease, and that customer service would increase beyond the 95% target.

Bottling Salad Dressing

A VSM proved to be particularly valuable to a salad dressing producing plant, by providing insight that prevented it from making a costly investment that would not have accomplished its objectives. The facility makes salad dressings, mayonnaise, and bottled sauces. On one particular salad dressing bottling line, market demand had increased Takt from the previous 300 bottles per minute (BPM) to 400 BPM. The plant manager's intuition, reinforced by some data, told him that the bottle-filling step was literally the bottleneck (this is one case where "bottleneck" is more than a figure of speech; the constraint to increasing line speed was the inability to get the fluid through the neck of the bottle at the desired rate). So he challenged his technical organization to increase bottle filling to 400 BPM. After some analysis, preliminary design, and prototyping, they informed the manager that with a redesign of the filling nozzles, the filling operation could indeed be elevated to 400 BPM. However, before proceeding with the investment in the new stainless steel nozzles, they asked me to develop a VSM to validate their flow rate possibilities. The VSM (Figure 19.3) clearly showed that even if the bottle filling machine were increased to 400 BPM, the line speed would increase to only 320 BPM. What no one realized until they collected the data to populate the VSM data boxes was that there were three other near-bottleneck steps that would become bottlenecks at the new Takt of 400 BPM. The homogenizer step in the kitchen area would reach its limit at 320 BPM, the label applicator at 360 BPM, and the case packer at 400 BPM. Because no one had ever tried to run at a rate above 300 BPM, the neighboring flow limitations had never been recognized. While it would be economically feasible to replace

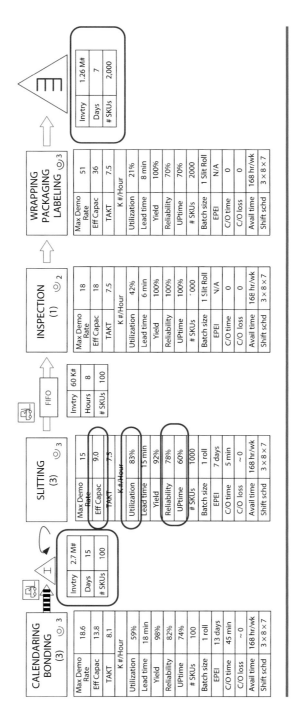

Figure 19.2 Sheet goods future state VSM.

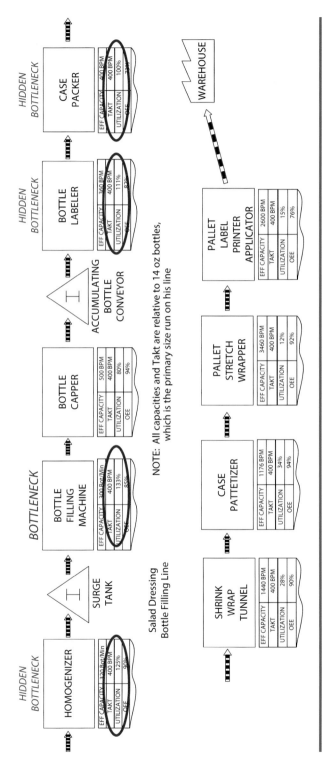

Figure 19.3   Partial VSM of a salad dressing bottling line (Takt and Utilization are at the target bottle rate).

nozzles on the bottle-filling machine, the total cost of eliminating all bottlenecks was prohibitive, so the decision was made to build a new line instead of upgrading the current one.

Had the technical team not decided to map the entire process, and had the hidden bottlenecks been overlooked, the plant manager would have committed a sizeable investment on the new nozzles only to discover later that the line was still 20% short of meeting Takt.

Cooling Towers in Polyethylene Production

A plant in Eastern Europe produces polyethylene pellets for use in manufacturing wrapping films, plastic toys, and other consumer goods. The operation extracts ethane gas from the ground, converts it to ethylene, and then polymerizes it into polyethylene, which is then pelletized. The process is exothermic and uses water to remove the heat from the process vessels and lines. The cooling water is cooled in large wooden cooling towers, which are more than 40 years old and in very poor condition. The cost of maintaining them and keeping them somewhat functional is high enough that they have a replacement program underway. A net-present-value analysis indicated that a 10-year plan would give the best return on the investment.

However, as we constructed the VSM of the polyethylene process, one of the control room operators pointed out that for much of the summer months, process flow had to be throttled back somewhat because the towers were not adequately cooling the water, so not enough of the process heat could be removed when running at normal flow rates. Thus, what had been viewed as primarily a maintenance problem now was seen as a productivity issue, a throughput limitation in a sold-out operation. This insight caused a re-examination of the replacement timing, and gave much more priority and urgency to the investment decision.

Producing Waxes for Coating Cardboard Boxes

A plant in Kentucky manufactures waxes and pastes. The primary customers for the wax products are corrugated manufacturers who use the wax to coat the stock from which cardboard boxes are made. The plant had two lines, A and B, which were believed to be completely independent of each other. The focus of my project was to design and implement a product wheel to optimize the sequencing and timing of the products made on line B. (A product wheel is a regularly repeating pattern of all the products made on a production line or major piece of process equipment. The sequence is optimized to reduce changeover difficulty and time, and the cycle is selected to balance agility and responsiveness goals with plant operability and OEE concerns. Product wheels are described in Appendix F.)

We began by developing a VSM of line B. The plant defines line B as beginning with a mixing step, fed from a resin polymerization step in another area of the plant, but we decided to begin the VSM at raw material input and include the resin polymerization, which feeds both line A and line B. We then decided to include line A on the VSM, even though it wasn't in the product wheel scope, to make sure we had a complete understanding of end-to-end flow.

Figure 19.4 is a block diagram of the plant. (The full VSM is not included here, to protect proprietary client process information.)

As we gathered all capacity and Takt data and began to populate the map, we discovered that there were some products run on line A that were heavy consumers of the resin, enough so that while those products were being produced, certain other high resin consumers couldn't be run on line B. In the current state, they had an informal process where they would at times delay certain products on one line to avoid conflict with the other line, but this wasn't documented and wasn't widely known.

With the discovery that this constraint existed, we decided to expand the program scope to include implementing a product wheel on line A. With the structure and the repeatable pattern that product wheels provide, we were able to coordinate the line A wheel with the line B wheel in a way that the heavy resin consumers on line B were offset in time from those on line A.

Thus, the VSM enabled us to get a broad understanding of flow and input material requirements and avoid a potentially serious bottleneck. It also led to a formal, documented, standardized process that resolved the constraint. The VSM allowed us to see the entire operation as an integrated system and therefore implement changes that benefitted the complete system and not just a major piece of it.

Improving a Capital Project Execution System

A mining and steel producing company in Eastern Europe had a capital project execution system that had very serious issues: cost overruns often exceeded 100%, and plant start-ups were sometimes as much as 24 months late. The company decided to completely redesign their project execution process based on Lean principles, to use Value Stream Mapping to understand where the defects and wastes in their current process were, and then create a set of future state VSMs to guide the redesign.

A current state VSM was developed, which highlighted the key wastes, time delays, and causes of rework, and then cross-functional process maps ("swim lane charts") were created to drive to root causes.

A lot of the mapping focused on the project Front End Loading (FEL) process. FEL is a term used to describe the pre-project planning activities to develop a conceptual design and ensure that it is aligned with business objectives and is practical to execute. It begins by drafting a Business Objectives Letter, to document and clarify what the business expects from the facility in terms of capacity, quality, lead time, product mix capability, and useful life. Conceptual designs

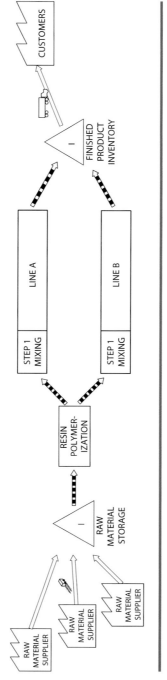

Figure 19.4 Block diagram of the wax production plant.

are generated, and process technology options studied. The process concepts and chosen technology are reviewed with all stakeholders, and once agreement is reached some general design can proceed so that cost appraisals and target schedules can be developed. The key feature of this approach is that the facility is conceived and designed in stages, beginning at a very high level and proceeding through more detailed levels, and that all appropriate viewpoints are engaged in approving each stage. Experience has shown that thorough FEL leads to fewer false starts and less redesign and rework as the project proceeds.

The VSM uncovered key gaps in the client FEL processes. Business objectives were not documented, so there was often confusion on what was expected from the facility. Key decisions on process technology selection were not made early enough in FEL, so too many technology alternatives were carried into the later stages of FEL, resulting in design costs for alternatives that would not be chosen. Input from key stakeholders and decision makers was missing in the early stages, resulting in a lot of rework later when their input was incorporated.

The mapping revealed several other causes of delays and overruns:

- There was confusion about roles and responsibilities of owner–operator engineers and full service design contractor engineers.
- There was no standard process for specification preparation, bid reviewing, or vendor selection.
- Cost libraries were far out of date, leading to inaccurate estimates.
- Excessive documentation and approvals were required (common in former Iron Curtain countries in Eastern Europe) causing significant delays.
- There was an intense focus on activities rather than on results (also somewhat common in this culture).
- Lessons from project failures were not being captured and analyzed for improvement on future projects.

Future state VSMs were created to rectify these problems. Future state swim lane charts were developed for each FEL stage to further clarify the specific improvements. Then detailed project management documents were written to capture the new execution process.

This is just one example of how VSMs can be used to guide the improvement of non-manufacturing, non-supply-chain processes. Other examples can be found in healthcare and banking.

Summary

Value Stream Mapping has proven to be an extremely beneficial tool in a wide variety of process operations, including food and beverage, batch chemicals, electronic materials, and paper and sheet goods. Its value has been demonstrated in the manufacturing portion of the process, in the finishing and packaging flow, and across the entire supply chain. Although the processes described in this chapter differ from each other, the benefits have a lot in common: greater process understanding, discovery of hidden waste, understanding of bottlenecks and flow constraints, and a roadmap toward a superior future state.

A VSM will not only be a strong enabler to improving your process, it can also provide insight that will help you avoid serious mistakes.

Appendix A: Process Industry Characteristics

Chapter 3 described some of the more significant differences between process and assembly operations, and that a VSM for a process operation should look a little different from one for a parts assembly process. The basic structure will be the same, with material flow, information flow, and a timeline, but the data boxes will require more data so that the wastes and the opportunities to improve flow and process performance can be more apparent. And the way parallel equipment is shown may differ to highlight additional flow improvement possibilities.

Taiicho Ohno recognized some of the differences and unique challenges of process operations as he was helping Toyota develop their production system. In *Toyota Production System—Beyond Large Scale Production,* he states:

> To be truthful, even at Toyota, it is very difficult to get the die pressing, resin molding, casting, and forging processes into a total production flow as streamlined as the flows in assembly or machine processing.

In short, he seems to be saying that the processes in Toyota that are similar to those found in the process industries are very difficult to make Lean. The key word is "difficult"; he does not say "impossible."

Characteristics That Distinguish the Process Industries

So, what are the differences that make it difficult to get into a streamlined flow?

Equipment Is Large and Difficult to Relocate

You can characterize many assembly processes by their relatively small, easy-to-move machines, such as drill presses, grinders, and lathes. On the other hand, most equipment in the process industries is very large and has process piping, hydraulic lines, and complex electrical wiring, making it very difficult to relocate to improve flow. Ovens used to dry extruded flakes that will become breakfast cereals can be 100 feet long or more, and 12 to 20 feet wide. A machine to form papers used as electrical insulation for power transformers can likewise have a footprint of 1000 square feet or more, and weigh several tons. Pieces of equipment of this scale are often referred to as "monuments" to emphasize their immobility. This makes it impractical to relocate equipment to improve flow, so concepts like cellular manufacturing require a broader view and a more creative approach. One such approach is described in Appendix C.

Processes Are Difficult to Stop and Restart

The machines typically found in assembly processes are easy to start and stop. This is not always the case in the process industries; process equipment is often time-consuming and costly to stop and restart. For example, chemical polymerization vessels may run a 2- to 4-hour batch that can't be interrupted without destroying the batch. Some continuous polymerization process lines typically run for months or years between shutdowns: stopping the flow of hot polymer causes the molten plastic to freeze in the lines. Prior to restarting, you would have to disassemble the entire process line and sandblast it clean or burn it out, at a cost of hundreds of thousands of dollars. Similarly, to stop a machine extruding plastic packaging films causes the molten plastic to freeze in the die lip, again requiring a very extensive and costly cleanup. This tends to drive long campaigns, overproduction, unnecessarily large inventories, and make the implementation of a pull replenishment system difficult. The push–pull interface concept described in Appendix D often provides a very practical compromise.

Capital Intensive vs. Labor Intensive

Chapter 3 described that process operations tend to be very capital intensive, where assembly processes tend to be very labor intensive. We know that labor productivity was a driving force in the development of the Toyota Production System. In Ohno's words:

> …This made the ratio between Japanese and American work forces 1-to-9. I still remember my surprise at hearing that it took nine Japanese to do the job of one American.
>
> Furthermore, the figure of one-eighth to one-ninth was an average value. If we compared the automobile industry, one of America's most advanced industries, the ratio would have been much different. But could an American really exert ten times more physical effort? Surely, Japanese people were wasting something. If we could eliminate the waste, productivity should rise by a factor of ten. This idea marked the start of the present Toyota production system.

And while labor productivity is an important factor in the process industries, asset productivity is typically far more important. As a result, a process VSM must describe the process in a way that highlights asset productivity and limitations so that waste can be seen and improvement possibilities understood. Lean programs should focus more on getting the most out of critical assets than on balancing labor tasks.

Hidden WIP

The Work In Process (WIP) in a process plant is often out of sight, and while not out of mind, often underestimated, so the inventory boxes on the VSM can be very surprising and enlightening.

In assembly plants, WIP often sits on the floor in bins or totes, on racks, or in carts where it is readily visible, whereas in process plants it is usually out of sight or hidden. In an architectural paint plant, the intermediate resins may be stored in 200- to 500-gallon stainless steel totes stored in an AS/RS (automatic storage/retrieval system, a high-rise rack storage system), far out of sight. Similarly, in a carpet manufacturing facility, master rolls of tufted, backed carpet, perhaps 12 feet wide and 6 feet in diameter, waiting to be dyed, will be stored in a large rack system, well removed from the tufting, backing, and dyeing equipment. What specific material is there, where it had been processed, and where it is headed next are difficult to determine without referring to a computer terminal connected to the inventory management system. So "walking the line" of a process plant to gather flow data for a VSM is not as effective as it is in the more visible assembly processes.

Product Differentiation Points

Chapter 3 defined "A" and "V" flow patterns and the increase in product types as material flows through a process plant, and why scheduling is critical at each differentiation point: if you make the wrong differentiation decisions, you will produce a currently unneeded variety, filling the warehouse with currently unneeded product. You have also consumed valuable production capacity, which is unavailable during that time to make needed products. Therefore, you end up with excessive finished product inventory, but ironically with poor customer service.

In a papermaking process, if the wide rolls of paper are slit to currently unneeded widths, finished product inventory of those widths will increase, while shortages of other widths can occur. A specific example of this is described in Chapter 19. In our synthetic fiber process, we have decisions to make in cut–bale about cut length, which if based on an imperfect forecast will unfortunately provide the opportunity to make products that are inappropriate to the current customer needs.

This profound difference in flow pattern, diverging vs. converging, provides both opportunities and challenges for the application of Lean tools. Fortunately, well-designed pull systems can facilitate better differentiation decisions at all divergence points. To avoid requiring the supermarkets in the later stages of the process to contain a managed inventory for each of the hundreds of types of semi-finished product, the ConWIP method described in Appendix D provides an attractive alternative to kanban. If lead times permit a finish-to-order strategy, it will eliminate the final stages of inventory where the highest product variety occurs and allow scheduling based on a forecast at the product family level, which is generally much more accurate.

Summary

Companies that manufacture toothpaste, house paint, salad dressings, synthetic fibers, carpeting, paper products, and bulk chemicals are all examples of what are called "process industries." Their manufacturing processes are significantly different from processes that make parts and assemble them into automobiles, refrigerators, cell phones, and bicycles. The type of equipment found in the various manufacturing steps of process industries is quite different from that found in parts assembly, and tends to be much larger, more expensive, and have a much higher impact on manufacturing performance and throughput. These factors make the application of many Lean concepts and tools more challenging.

The most important differences between assembly and process operations, and how Lean thinking can be applied, are cataloged in Table A.1.

Perhaps the greatest distinction between process lines and parts assembly is the high degree of product differentiation that occurs as material moves through process operations. Assembled products are usually made from tens, hundreds, or thousands of component parts, which result in a small number of finished product types. Process operations, on the other hand, usually start with a small number of raw materials, from which hundreds or thousands of final products are made. This difference in flow pattern—divergence vs. convergence—has a profound impact on flow dynamics, and requires appropriately designed pull systems to ensure that differentiating decisions are made in a way that results in products that meet current needs while avoiding products that don't. A well-constructed VSM provides a clear picture of the flow patterns and the data required to develop a pull strategy appropriate to your specific operation.

Table A.1

Industry	Assembly Manufacturing	Process Industries
Examples	Automobiles Aircraft Cell phones Computers Power tools Industrial equipment Home appliances	Paints Batch chemicals Paper and plastic sheet goods Food and beverages Personal care: shampoo, toothpaste Carpets Metals and ceramics
Process Flow Model	"A" processes • Part variety convergence • Many raw materials • Little final differentiation	"V" processes • Material variety divergence • Few raw materials • High final differentiation
Prime Influence and Guidance	Toyota Production System	TPS with extensive application experience
Primary Focus	Waste • Inventory reduction • Overproduction • Defects	Waste, cycle time • Inventory reduction • Throughput • Yield losses
Primary Economic Drivers	Labor productivity Inventory reduction	Asset productivity Inventory reduction Increased throughput Reduced yield losses
Other Major Focus Areas	Flow Level production to Takt Lot size of one	Flow Level production to Takt Lot size optimized by equipment size
Primary Rate Limiting Factor	Labor	Equipment
Tools/ Techniques	Value stream mapping 5S Standard work Poke-yoke SMED One piece flow Cellular manufacturing Production leveling Mixed model production Autonomation Synchronize flow to Takt Pull systems	Supply chain mapping Value stream mapping 5S Standard work Poke-yoke SMED Flow determined by equipment size Cellular manufacturing Product wheels Autonomation Synchronize flow to Takt Pull systems

Continued

Table A.1 (Continued)

Industry	Assembly Manufacturing	Process Industries
Batch logic influenced by	Machine set-up time Transportation lot size	Batch size by: • Equipment size and characteristics Campaign size by: • Changeover time • EOQ
Set-Up Issues	Time to replace, reset tooling	Time to clean out process vessels Material losses flushing lines out Time to reset, stabilize temperatures Time for pressures to equilibrate Time to get properties on aim after restart Material losses getting properties on aim
Drivers for Cellular Manufacturing	One-piece flow Labor utilization Flow visibility Flow management Reduced WIP Facilitate pull	Simpler changeovers Asset utilization Flow management Yield improvement Reduced WIP Facilitate pull
Cellular Manufacturing Implementation	Group technology Physical work cells	Group technology Virtual cells
Production Leveling Techniques	Control market demand Mixed model production Heijunka	Product wheels: • Campaign sequence optimization • Campaign length optimization • Heijunka

Appendix B: SMED Principles

The current state VSM generally shows that some or all inventories are higher than they really need to be. And while one frequent cause is that inventories have been set by tradition or gut feel rather than by sound mathematical principles, another equally prevalent cause is that the campaign cycles (EPEI, "Every Part Every Interval" in Lean terminology) on key pieces of process equipment are very long. The root cause of these long campaign cycles is that changeovers are difficult, time consuming, and expensive. The natural reaction to that is to avoid these difficult changeovers by running long campaigns which has the unfortunate consequence of driving inventory higher.

Therefore, if the current state VSM shows high inventory levels, the move to an improved future state must place a heavy emphasis on changeover improvement. SMED (Single Minute Exchange of Dies), conceived by Shigeo Shingo while working at Toyota in the 1950s and then perfected by Toyota over the next 30 years, has become a widely used best practice for changeover simplification and reduction.

SMED Origins

As the Toyota Production System was beginning to evolve in the early 1950s, Toyota recognized that it was critical that product changes be accomplished as quickly as possible so that short campaigns would be feasible.

One of the most time-consuming changeovers they faced was the replacement of the dies on the large hydraulic presses used to stamp out auto body parts. Shigeo Shingo, an Industrial Engineer consulting with Toyota, developed a methodology for examining all set-up operations and modifying the set-up process to reduce the overall time. Using Shingo's techniques, Toyota was able to shorten the die changes from 3 hours in the 1950s to 15 minutes by 1962, and to an average of 3 minutes by 1971 (Ohno, 1988). In recognition of this tremendous accomplishment, Shingo's methods and techniques have become the standard for changeover reduction and have come to be known by the acronym SMED, for Single Minute Exchange of Dies.

SMED Concepts

Four of the fundamental ideas that SMED promotes are (Figure B.1)

1. Determine if any of the tasks done during the changeover can be done before the equipment is turned off and production stopped or after the equipment is turned back on and making good product. These tasks can consume a lot of time, so moving them outside of the time window when the machine is not producing can shorten changeover time dramatically. Any tasks that must be done during the changeover are called "internal" tasks, while those that could be done before or after are called "external" tasks. So step 1 is to identify any external tasks.
2. Move external tasks outside of the changeover time.
3. Simplify the remaining internal tasks.
4. Where feasible, perform internal tasks in parallel. If several operators can perform tasks concurrently, the time can be reduced without increasing the total labor content of the set-up.

After the changeover process has been revised and tested, it is critical that it be documented, standardized, and audited on an ongoing basis so that the improvements can be sustained.

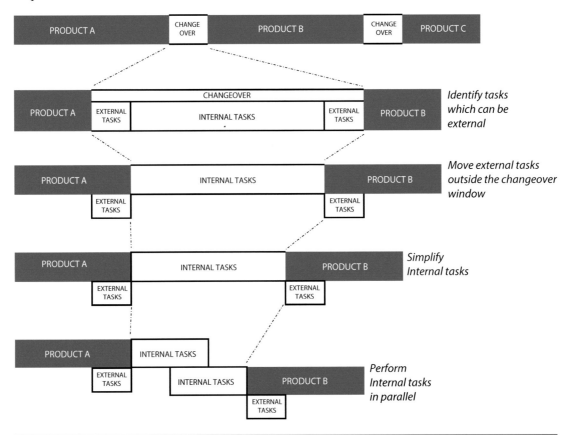

Figure B.1   The SMED changeover improvement process.

It is also critical that after the improvement is demonstrated, the campaign cycles are revised to take advantage of the changeover time reduction and that inventory targets are reset in accordance with the new cycle.

Product Changeovers in the Process Industries

SMED is particularly valuable on process industry operations because the changeovers tend to be much more complex and involve more costly material losses.

In process operations, a lot of time may be spent in cleaning out the raw material feed systems and the processing equipment to prevent cross-contamination. For example, the tinting tanks used in paint manufacturing require thorough cleaning during color changes. In many food processing plants, equipment is shared among several product varieties, which may or may not contain allergens such as peanuts. This can pose very stringent requirements for cleaning between products, followed by extensive testing to ensure that the equipment is free of contaminants. The cleanouts and testing required in pharmaceutical manufacturing are even more thorough. The tasks performed during these cleanups are very well suited to SMED analysis.

In extrusion, sheet good, and batch chemical processes, much of the time lost is the time required to bring the line to the appropriate temperature, pressure, speed, thickness, etc. after all the mechanical tasks have been performed. Therefore, the SMED process is very helpful in attacking these so that the total time for the changeover, including the time for process conditions to stabilize, is reduced. For example, pre-heating parts that must be installed during the changeover can make the majority of the pre-heat an external task.

Summary

If time and material losses during changeovers can be reduced, overall production cycles can be shortened, which will enable reductions in inventories. And because inventories in most process operations are higher than they need to be, and are frequently the most significant waste found on the VSM, moving from the current state to an improved future state frequently requires a changeover improvement process like SMED.

SMED is not a one-time event. Experience in many applications has shown that repeating the process periodically will surface opportunities that were not found previously. Toyota's experience is a dramatic example: it took them 20 years of relentlessly repeating the SMED process, but as a result, they went from 3 hours to 3 minutes!

Appendix C: Cellular Flow

Of all of the improvement tools in the Lean toolbox, Cellular Manufacturing is perhaps one of the most powerful. It enables smaller lot production, more visible flow, quality improvements, reduced Work In Process (WIP), and shorter lead times.

Any time the current state VSM shows that there are several pieces of equipment in parallel at each step, the process is a candidate for cellular flow. The cellular concept separates material flow paths and product groupings into much more manageable subsets, where each cell has fewer products to process, and those products are divided into families with similar processing characteristics so that changeovers are simpler and shorter.

Typical Process Plant Equipment Configurations

Cellular manufacturing only applies where you have similar equipment in parallel at one or more points in the process. It is particularly beneficial where there is parallel equipment at several steps, like the configuration shown in Figure C.1, a pattern typical of many process industry plants. There are a small number of key processing steps, in this case four, and there are a few (three or more) machines, tanks, or reaction vessels in parallel at each step. The parallel machines are quite similar, and often a specific material can be processed by any one of them. Occasionally, the machines or vessels have some unique capabilities such that some materials must go to a specific machine or vessel.

Figure 2.1 showed this configuration for the synthetic fiber process. There are three identical polymer reactors, six similar spinning machines, five draw–anneal machines, and four balers. Flake produced on any of the three reactors may be processed by any of the spinning machines, draw–annealers, and balers.

Process plants usually require this array of equipment to handle the high volume of material to be produced. Practical equipment size limitations prohibit the design of a single machine or vessel large enough to process the full throughput required. Product mix considerations, i.e., the high degree of product variety, would encourage the use of many small machines to give

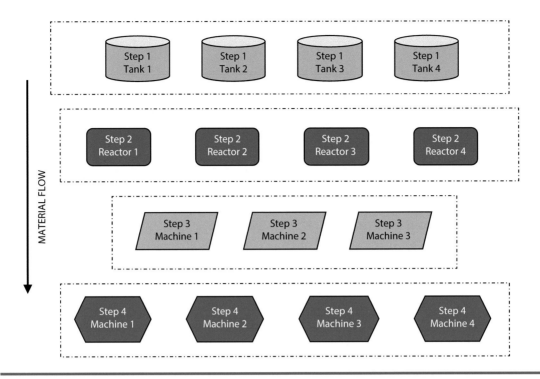

Figure C.1 **Typical process industry equipment footprint—functional configuration.**

more flexibility, but economies of scale have deterred plant designers from going in that direction. As capital cost optimization has traditionally overridden good Lean thinking in process plant design, the result is a few large vessels or machines at each process step.

This equipment configuration is highly valued for the flexibility it offers. If a batch of material is leaving step 1, and one of the step 2 machines is down for maintenance, there may be others available to process the material. The result is that flow paths are often as shown in Figure C.2. All of the flexibility inherent in the system is exploited, but generally with more negative than positive consequences. There is frequently a belief that utilizing this flexibility maximizes asset utilization, although the opposite is usually true.

This mode of operation, taking advantage of the inherent flexibility of this configuration, brings a number of problems. Material tends not to flow directly from one step to the next, but to be put into some type of storage. In the synthetic fiber process, flake does not flow directly to a spinning machine; instead, it is stored in the silos, to be conveyed later to a spinning machine. Thus, large WIP storages are created. Flow becomes unsynchronized, is difficult to visualize, and is even more difficult to manage. Because each piece of equipment can process any of the product types, each may have most of the products from the entire product lineup on its schedule.

Quality suffers for two reasons. There is a significant time lapse between each process step, so any defects or out-of-spec material may not be discovered for some time, making all of the intermediate material suspect. Even with this simple-looking arrangement, there are 192 (4 × 4 × 3 × 4) possible flow path

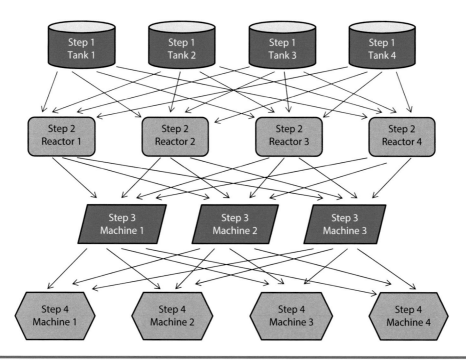

Figure C.2   Typical process industry flow patterns.

combinations. Since no two machines or vessels will produce exactly the same product, it provides 192 different ways that process variability can add up. A Statistical Process Control (SPC) specialist would tell you that you don't have *a* process; you have *192 different* processes. With so many variables, root cause analysis of product defects becomes very difficult.

Because there are thought to be alternate paths available whenever a piece of equipment fails, there is far less urgency to maintain the equipment appropriately. Thus, with time, equipment performance as measured by OEE deteriorates. We saw this with the sheet slitters described in Chapter 19.

Cellular Manufacturing Applied to Process Lines

In discrete parts manufacturing processes, cellular manufacturing has traditionally required relocating the equipment into U-shaped or L-shaped patterns, to provide shorter paths for operators to travel. It has therefore been thought that cells were not applicable to most process operations because the equipment is most often very large and expensive to relocate. But the advantages of cellular flow can be achieved without moving anything, by creating virtual flow paths, so that flow is managed in a cellular fashion. The key is to think in terms of flow rather than function.

The basic concept is straightforward: start by grouping all process materials or products into families requiring similar process conditions. Then identify the process equipment required by each family, but instead of creating a work cell by rearranging the equipment, create virtual work cells by defining the

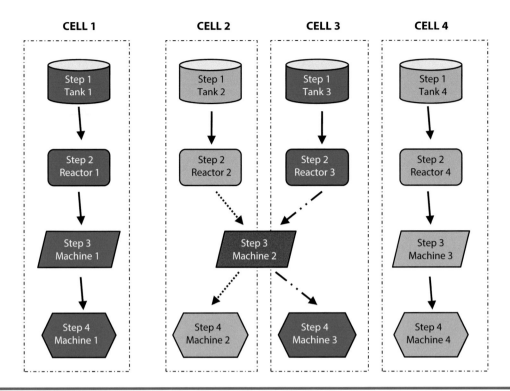

Figure C.3 Grouping into virtual work cells.

acceptable flow patterns. Figure C.3 shows what this would look like for the process diagrammed in Figures C.1 and C.2. Again, no equipment would have to be moved; the new, more limited flow patterns would simply have to be defined and followed.

The advantages of this virtual work cell concept are:

- Flow becomes far easier to understand, to visualize, and to synchronize.
- Flow tends to be more continuous, with less material being transported to storage, so WIP and material handling are reduced.
- Quality improves because feedback is immediate.
- As depicted in Figure C.3, we now have only four possible flow paths instead of the 192 we had before, so product variability is dramatically reduced.
- Each cell is generally processing a subset of the full product mix with similar requirements, so changeovers become far easier and usable capacity increases.

It must be recognized that the numbers don't always work out perfectly, but that reasonable compromises can usually be found which will give most of the benefit. In the case shown, because there are only three machines at step 3, one must be shared between cells 2 and 3. If that machine didn't have enough capacity to process the total throughput of the two cells, additional compromises would have to be made, perhaps as shown in Figure C.4, with six possible flow path combinations. But, comparing Figure C.4 to Figure C.2, you can see that even in the worst case, flow paths will be far reduced from the non-cellular flow.

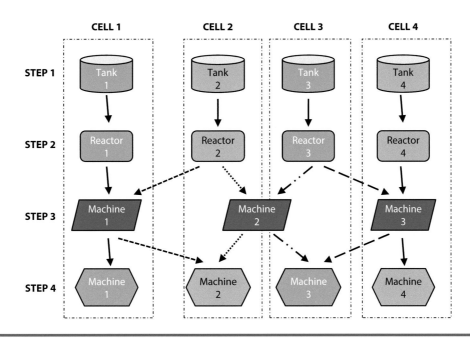

Figure C.4 Alternative virtual cell grouping.

Figure 15.3 showed a virtual cell layout for our synthetic fiber process. Because there are only two polymer reactors in this stage of the future state, reactor 1 would be shared between cells 1 and 2, i.e., spinning machines 1, 2, and 3. With four steam annealers, each will take the output from two spinning machines. But the flow paths are fixed as shown: steam annealer 1 will take material from spinning 1 and 2 only, for example. The flow is now restricted to the eight possible combinations shown in Figure 15.3, where prior to the formation of virtual cells, 192 possible paths ($2 \times 6 \times 4 \times 4$) existed. Thus, the key advantages of cellular manufacturing apply.

Although the new flow paths are the most readily apparent aspect of cellular manufacturing, the more significant benefit is that the product portfolio can be divided up into families, so that each piece of equipment must process far fewer product types. Having fewer products on each cell dramatically simplifies flow and pull.

An even more profound benefit is that the products assigned to a specific piece of equipment are usually grouped by similar processing characteristics, so that the changeovers on a piece of equipment are far faster and less expensive. For example, if the products assigned to a specific spinning machine are all spun at the same filament diameter, the time required to change spinnerettes on product changeovers is gone.

The part of the cellular manufacturing process where products are grouped into families with like characteristics and processing requirements is often called Group Technology. The APICS dictionary defines group technology as "an engineering and manufacturing philosophy that identifies the physical similarity of parts (common routing) and establishes their effective production. It…facilitates a cellular layout." Even if the equipment can't be arranged into virtual cells, or if there is parallel equipment at only one step, Group Technology still offers a significant advantage for that parallel equipment, by dividing the product lineup so that each of those parallel pieces of equipment has a smaller product mix to process.

Summary

Where the equipment configuration lends itself to cellular manufacturing, it should be considered for incorporation into the future state. The improved flow patterns resulting from application of cellular concepts reduce several kinds of waste as well as improving flow. Where the configuration does not lend itself to cells, the concept of Group Technology should be applied wherever there is similar equipment in parallel. Not only does this reduce the number of products on each piece of equipment, but a grouping can usually be found that reduces changeover time and cost, thus enabling shorter campaign cycles and lower inventories.

Therefore, wherever the VSM shows parallel equipment at a step, Group Technology should be considered, and if parallel equipment exists at several steps, dividing it into virtual cells should be investigated.

Appendix D: Pull Replenishment Systems

Why Is Pull Important?

The largest waste found on most process VSMs is inventory waste. We have described a number of ways to reduce this waste: SMED to reduce changeover time so that campaign cycles can be shortened, thus allowing inventory reduction, cellular manufacturing to streamline flow and dedicate products in a way that shortens changeovers and thus campaign cycles, and product wheels to further optimize campaign cycles. Another tool that offers additional improvement beyond those is the pull replenishment concept, where production is scheduled not from any forecast but is based on the current status of orders and inventories.

Pull is such a fundamental part of Lean systems that it almost always appears on future state VSMs.

What Is Pull?

The pull concept has two main components:

1. Production is allowed only to replace inventory that has been consumed ("pulled" from inventory) since the last production of this material, or to satisfy immediate orders.
2. The amount needed is communicated through visual signals, which are readily accessible to everyone needing to understand the current and upcoming production requirements.

Thus, pull includes both replenishment of consumed stock and production in response to a signal from a downstream operation that more material is needed now. Art Smalley, in *Creating Level Pull,* uses the terms "replenishment pull" and "sequential pull" to distinguish them.

A pull replenishment system is a significant improvement over the more traditional "push" system where production is scheduled based on forecasts without considering current inventories, and material is then "pushed" through the process based on these forecasts. Push very often leads to overproduction and high inventories. In cases where consumption has exceeded the forecast, push can lead to stockouts.

The objectives that Taiichi Ohno had in mind when he conceived pull for Toyota were to synchronize flow with Takt, to reduce inventory waste, and to reduce the material management tasks associated with production planning and scheduling by implementing visual signals to inform everyone what was currently needed.

Pull in Assembly

The vast majority of documented cases of pull production are in assembly processes. Pull has brought great value to these processes for the very reasons that Ohno and his contemporaries at Toyota developed it: It reduces or eliminates overproduction, synchronizes flow to current demand, and simplifies all the managerial tasks associated with production scheduling.

Many of these applications employ kanban logic, which forms the communication chain flowing in reverse through the process, starting with customers, and flowing back all the way to raw materials. Kanban literally means a signal, or a visible sign. In typical systems, the signal can take the form of cards, totes, or spaces marked on the floor that signal need to replenish when empty.

The basic kanban concept is shown in Figure D.1 and Figure D.2. In this example, it has been determined that four containers of finished product is the appropriate inventory to satisfy customer demand for one specific SKU in a timely fashion, so the finished product supermarket has four containers with that product. (Note that the term "supermarket" is often used to describe inventories managed by kanban principles. Indeed, Ohno first conceived pull and kanban processes after studying American supermarkets in the early 1950s. He was

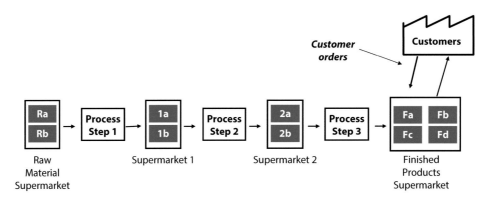

Figure D.1   Kanban concept.

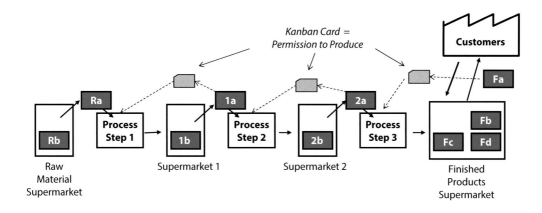

Figure D.2 Kanban operation.

impressed that in a supermarket, customers can get what they need, in the quantity they need, when they need it, and that the shelves were restocked with only what customers had pulled from the shelf.)

It has also been determined that two containers of WIP are required between each process step to ensure smooth flow through the plant. When a customer orders a container (see Figure D.2), one container (Fa) is removed from inventory and shipped to that customer. Before it is shipped, a card on the container is removed and sent to the final processing step (Step 3) to signal permission for that step to produce another container of that material to replace the one the customer has pulled. In order to produce a replacement, Step 3 needs input material, so it pulls a container of WIP (2a) from Supermarket 2. A card on that container is removed and sent to Step 2 to signal permission to replenish. In that manner, as material is moving from left to right through the process, the kanban signals, in the form of cards, are moving from right to left, or upstream in the manufacturing process.

Instead of cards, this system could use empty containers as the kanban signals. When the material in container Fa is shipped to the customer, the empty container could be sent back to process Step 3 to signal a need to replenish. After Step 3 has pulled WIP Container 2a and consumed the material, that container would be sent back to Step 2 to permit production to refill it. Another alternative is that the system could use empty container spaces in the supermarket to signal that a container has been pulled and needs to be replaced.

Some key points about pull, kanbans, and supermarkets are:

■ Although inventory is waste, some inventory is needed by most processes to maintain flow that matches Takt.
■ In a pull system, inventories are managed in a way that prevents overproduction and thus limits inventory to the minimum required for smooth flow.
■ Kanban signals are one very effective way to manage flow and inventories.
■ Kanbans are not the only way to achieve pull. An alternative, called ConWIP, will be described later.

- In any case, there must be a communication or signaling process that is visible to all who are involved in real time replenishment decisions, that is, anyone who gives or receives permission to produce. The signals can be in the form of cards or containers, or can be electronic signals triggered by an inventory management or production scheduling system.
- Production is triggered either by consumption of material in inventory or by a demand signal from the downstream operations, not by a schedule created from a forecast.

Difficulties in Process Plants

Process plants pose two unique challenges that must be met for pull to be feasible. The first challenge is that pull inherently requires that equipment be stopped and started within short periods of time. When a given step in a process has produced enough to meet the immediate need, whether signaled by kanban or by some other signal means, the step must stop or material will be produced that is not currently needed. This is one of the most fundamental characteristics of any pull system—production stops as soon as the need has been satisfied. A difficulty arises when production equipment cannot be stopped, which is not uncommon in the process industries. Many plastic pellet, synthetic fiber, and paint manufacturing operations begin with a polymerization process, which takes various chemicals, mixes them, and them reacts them at high temperatures and pressures, so that the molecular chain length will grow and properties such as tensile strength and viscosity will increase. Continuous polymerization processes are very difficult and costly to stop. In many cases, if flow stops even for minutes, the polymer will solidify in the vessels and piping, requiring that everything be disassembled and burned out before the process can be restarted. (That's the reason that the spinning machines in the Riverside Fiber process continue to have polymer flowing during a product changeover.) It is not uncommon for these processes to run for 24 to 30 months before stopping for a process overhaul.

A second challenge when applying pull in process plants is the large number of product types in the latter stages of production. If inventories are managed by kanban in a process with 200 semi-finished types, there must be kanbans for all 200 types, which can be a large amount of inventory, thus defeating the primary purpose of pull. Fortunately, there are techniques for addressing both of these challenges.

Push–Pull Interface

When designing a pull system for a process with a step that can't readily be stopped, creating a push–pull interface is often a reasonable compromise. In many cases, this offers much of the benefit of pull, without incurring the extreme cost of stopping the challenging pieces of equipment. The concept is to find a point in the process downstream of all difficult-to-stop steps where inventory can conveniently be maintained, and designate it as the push–pull interface. Material is then pushed through the earlier process steps to accumulate in the inventory at the push–pull interface. Material is then pulled from this inventory only as downstream needs are signaled.

This works well in many process operations because the steps with difficulty in stopping are usually early in the process, where little product type differentiation has taken place. Forecasts are generally more accurate at this aggregate level, so the mix can be adjusted so that the materials being pushed are likely to be the specific types currently needed. In other words, if you must overproduce, it is better to overproduce materials that are likely to be needed sooner rather than later. At the downstream steps, where most of the differentiation typically occurs, material is being pulled, so the product types being made are those needed immediately. Figure D.3 illustrates the concept for a pellet extrusion process, where the polymerization and extrusion steps are very difficult to stop, but the blending and compounding extrusion steps can be interrupted more easily.

ConWIP

ConWIP (CONstant Work In Process) is a strategy for managing production in a way that limits in-process inventory and can be used as an alternative to kanban. ConWIP can control an entire process, or several steps in the process. The fundamental idea is that nothing is allowed to enter the process segment being controlled by ConWIP, the ConWIP loop, until something leaves. This one-for-one relationship between lots leaving and lots entering maintains WIP within the loop to a fixed amount. Thus, overproduction is prevented, so ConWIP can be used as the control mechanism in a pull system.

One of the most significant features of ConWIP is that the only material in any WIP storage within the process is in the materials currently being processed. In a process industry plant where hundreds of final SKUs can be produced, the WIP within the operation at any time is only in a fraction of those. In contrast, in a kanban-controlled system, WIP must include kanban quantities for all product types that exist at that point. Thus, ConWIP can overcome the process industry challenge mentioned earlier, by requiring only reasonable amounts of WIP even in highly differentiating processes. For that reason, ConWIP has sometimes been called part-generic kanban in contrast with the more traditional idea of part-specific kanban. The concept is illustrated in Figure D.4

Visual Signals

Visual signals are an extremely important part of any pull system. Kanban, the heart of Toyota's pull system concept, literally means visible sign. Although card

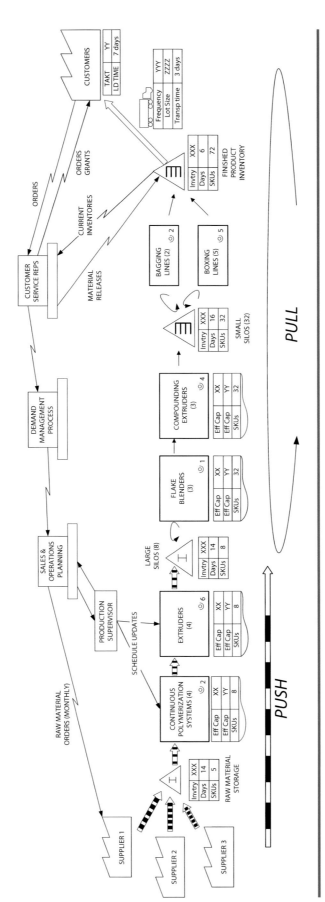

Figure D.3 Push–pull interface example.

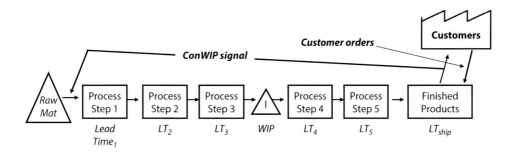

Figure D.4 ConWIP concept.

systems are often the signal of choice in assembly operations, and have seen successful application in some process operations, visual scheduling boards called Takt boards have proven to be a very effective way to satisfy this need. Takt boards have seen widespread use in process plants in a variety of different types of processes, ranging from the manufacture of plastic pellets to the production of brake fluids.

To get a more detailed understanding of how Takt boards provide the visual communication, consider a process line making plastic materials that customers mold into gears. The line runs a 7-day product wheel (see Appendix F), which begins every Wednesday morning at 8 a.m. On Tuesday morning, the production planner checks the current inventories of all materials made on that line. From that, she determines what must be made on the next wheel cycle to return inventories to the cycle stock plus safety-stock target. What the production planner will schedule to be produced will typically not be exactly the cycle stock. If the inventory data show that some of the safety stock has been consumed on the current cycle, she will schedule replacement of that safety stock in addition to the cycle stock. On the other hand, if not all of the cycle stock has been consumed, and there is more than the safety stock remaining, she will schedule production of less than the cycle stock. The fact that there is an offset in time from the day on which the inventory is checked (Tuesday) and production of a specific material will be started (say, Friday) constitutes what some operations management texts call a "review period." This can be easily accommodated, and is covered in Appendix E.

Once the quantities to be produced have been calculated, the production plan is set. The sequence in which the various products are to be made had been determined as part of the product wheel design. The production plan will then be communicated to the person responsible for running the line. This may be an area supervisor, or may be a flow manager. In this case, communication of the plan is done electronically because the planner is located in an office several hundred miles from the plant.

The flow manager checks the plan to make sure that it accommodates any special events, such as production of a test product or a planned outage. He then meets with a member of the production team, who puts the schedule on the Takt board, which in this case is handwritten but could be electronic. Thus, the board provides a visual signal to everyone in the area as to what is to be produced to replenish what has been consumed from inventory. The Takt board has the added benefit of providing a place to record production to the plan and to log production difficulties. Visual signals to support pull are discussed in much more detail in King (2009).

When to Start Pulling: The Sequence of Implementation

There are at least two schools of thought on when it is most appropriate to begin to implement pull on part or all of a process. One school believes that getting to pull is an essential component of any Lean journey, and that pull is the platform to drive continuous improvement. Therefore, pull should be implemented soon, and it will then drive necessary process flow improvements.

The more appropriate concept (Figure D.5) is that pull is better done after other flow improvements have been accomplished, after the process is flowing smoothly, with high stability and discipline, and all process variability has been reduced as much as possible. The concept is that pull will be easier to design and install if the process has been simplified and stabilized, and will be easier to sustain if there are fewer interruptions and special cause events.

The latter view is reinforced by Umble and Srikanth in *Synchronous Manufacturing* (1990):

> One weakness of the JIT approach is the inability to identify systematically the critical capacity constraint resources in the operation in advance. The Japanese approach to attacking waste and supporting the process of continuous improvement within the plant is essentially unfocused.

Value Stream Focus

It must be emphasized that what is being pulled in a pull system is the value stream, not the equipment. Lean implementers sometimes say that they want to

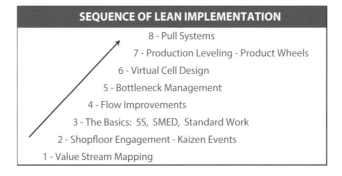

Figure D.5   Preferred path to pull.

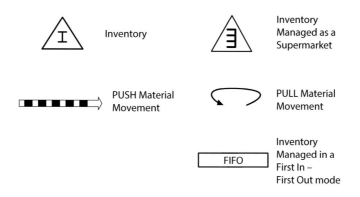

Figure D.6 **VSM icons for pull systems.**

put an extruder, an autoclave, or a carpet-tufting machine on pull. What they mean to be saying (or should mean) is that they want to put the value stream flowing through that piece of equipment on pull. This is an important distinction, and far more than just semantics or terminology.

Showing Pull on a Value Stream Map

Figure D.6 shows the VSM icons for supermarket inventory and flow when pulling from a supermarket.

A future state VSM incorporating pull is shown in Figure 15.4.

Summary

A pull system is one in which day-to-day production is based on current conditions on the plant floor rather than scheduled from a forecast, production is synchronized to true customer demand, and inventory is kept to the minimum needed for smooth flow. Although the term pull may imply that we are replacing material that has been pulled from storage ("replenishment pull"), pull can be used to produce material needed to fill current customer orders as well ("sequential pull").

The traditional way to manage pull in parts manufacture and assembly uses kanban signals in the form of cards, containers, and available space in a storage area. Although kanban is sometimes used in process plants, the high number of product types often favors an alternate system called ConWIP (also called part-generic kanban) vs. the more traditional part-specific kanban.

Pull systems inherently require that manufacturing equipment be started and stopped in accordance with kanban or ConWIP signals. With some process equipment, stopping in this way is not practical because a shutdown and restart can cost hundreds of thousands of dollars. Consequently, many process lines make use of a push–pull interface, where material is pushed through the difficult-to-stop equipment to an in-process inventory. Material is then pulled from that inventory as needed to feed downstream operations.

Although pull can offer great benefit to process lines as well as assembly lines, you should not begin to design pull until the process has been stabilized and variation has been reduced as much as possible. If virtual cells are appropriate, they should be implemented before pull, as should product wheels. With all the foundational improvements successfully in place, you are then well prepared for a very successful and effective pull implementation.

Appendix E: Cycle Stock and Safety Stock

We've talked a lot about the fact that inventory is one of the most pervasive wastes in process operations, and have suggested a number of Lean concepts that can improve process performance to reduce that waste. But the first thing that should be done is to understand if you really need the inventory you currently carry, if the current process performance requires that much. Very often, the inventory is more than needed even under the current conditions, so before trying to improve performance you should right-size the inventory based on current performance. This is generally a zero or low cost way to reduce inventory waste before putting effort into the specific Lean approaches to inventory reduction, such as SMED, virtual cells, product wheels, and pull systems.

   If the rule of thumb discussed in the section on inventory data boxes in Chapter 9 tells you that the inventory you see on the VSM is high, there's a good possibility that you can make a significant dent in it simply by performing these calculations and then letting the inventory naturally fall to the appropriate levels, i.e., stop buying or making the specific high SKUs until the normal demand takes them down to the appropriate level.

Cycle Stock and Safety Stock

Where inventory is deliberately being maintained, it generally has two components—cycle stock and safety stock. Cycle stock is the inventory carried to accommodate the cyclic nature of material delivery or production. It is the amount of a specific product to be made during the production cycle, to satisfy demand over the full cycle including the portion of the cycle when other products are utilizing the asset. For example, if the production process were based on a 7-day product wheel, the cycle stock for material A would be seven days. If material A occupies one day on the wheel, at the end of its production day there must be six days of material in the next downstream process step or in the next inventory location. That material is needed to satisfy demand for material A in the 6-day interim until material A will be made again. Therefore, the cycle

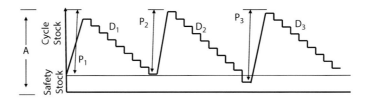

Figure E.1 Cycle stock and safety stock.

stock for material A includes the one day that was consumed while material A was being produced, and the six days to satisfy demand during the rest of the cycle.

The second component of inventory is safety stock, material held to satisfy demand in cases where actual demand is higher than expected, or where the next cycle is late in starting.

Figure E.1 shows a profile of inventory vs. time for a single SKU in a case where cycle stock and safety stock are present. In production period P1, cycle stock is produced to raise the level to A. Demand during the next cycle, D1, is equal to the average demand, so the cycle stock is consumed but safety stock is not. Production P2 raises total inventory back to level A. Demand during the next cycle, D2, is higher than average, so in addition to consuming all the cycle stock some of the safety stock is needed. This would also be the case if it took longer than average for the process to complete its cycle and return to making this material. Thus, the safety stock will protect flow against either variation in demand or variation in supply lead time. Production P3 must be greater than average to replace cycle stock plus the amount of safety stock that was consumed.

Cycle stock is based on the average demand expected. This can be based either on demand history or on a forecast. If previous demand is considered the best predictor of future demand, demand history should be used to set cycle stock. If there is a forecast that is believed to be a more accurate indication of future demand, cycle stock should be based on the forecast. Because forecasts can vary period by period, the cycle stock may be adjusted upward or downward each period in accordance with the forecast.

Some may consider that carrying safety stock is counter to Lean, that it is waste, that it is material being stocked "just in case." Although it is waste, it is necessary to protect customers against the variations in our process and the variation with which the customers themselves present us. Until these variations can be eliminated, smooth flow of products to customers requires some level of safety stock.

Safety stock is being carried because we intend to use it, and will use it frequently. Because demand will be higher than average about half the time and lower than average the other half, you should expect to consume some of the safety stock during about half of the cycles, as depicted in Figure E.2. Further, if safety stock is calculated based on a 95% cycle service level, stockouts should be expected on 5% of the cycles.

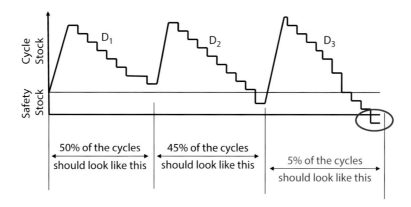

Figure E.2   With a 95% service level, stockouts can happen.

Calculating Cycle Stock

Cycle stock can be replenished on a fixed interval or on a fixed quantity basis. (These are the two major models; there are others, which will not be discussed here.) As the name implies, fixed interval replenishments occur on a regularly repeating cycle, where the time between replenishments may change slightly, but the quantity can vary significantly, depending on how much material has been consumed during the most recent cycle. Fixed quantity replenishment behaves the opposite way: The quantity is determined based on some specific criteria and doesn't vary. The interval can vary significantly, again based on the rate of consumption since the last replenishment.

In our synthetic fiber process, raw material inventory is replenished based on a fixed quantity model, where the fixed quantity is a railcar. All other inventories are refilled based on a defined production cycle, or EPEI, so they are fixed interval.

Fixed Interval Replenishment Model

Figure E.3 shows the inventory profile for a single material in a fixed interval strategy. The specific case shown is a product wheel, with a 14-day wheel time, but it could also depict a process step with a 14-day EPEI scheduled by some other strategy. In producing this material, we must make enough to last until the

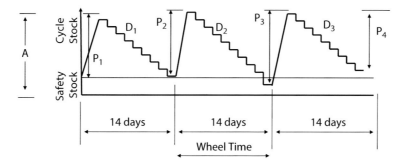

Figure E.3   Inventory profile with fixed interval replenishment.

next production of this material, or 14 days' worth. Therefore, the cycle stock will be the average demand during a 14-day period, and the peak inventory will be cycle stock plus safety stock, or 14 days' worth plus safety stock. What actually is produced during any cycle is not always the cycle stock, but depends on current inventory. In Period 2, for example, demand D2 is slightly greater than average, so some of the safety stock has been consumed. Thus, the quantity to be produced, P3, will include the normal cycle stock plus the amount of safety stock that was consumed.

The equations that apply to those situations are as follows:

$$Peak\ Inventory = Cycle\ Stock\left(1 - \frac{D}{PR}\right) + Safety\ Stock$$

$$Average\ Inventory = \frac{1}{2}(Cycle\ Stock)\left(1 - \frac{D}{PR}\right) + Safety\ Stock$$

where D is the demand for that material per unit of time, and PR is the production rate, the total quantity produced over that same time.

The $(1 - D/R)$ term is there to reflect the fact that some of the cycle stock is being consumed by downstream steps during its production cycle. This has a minor effect on products with a relatively small portion of the production cycle, but can be significant if a product occupies a large portion of the cycle.

The quantity to be produced on any cycle will be:

$$Quantity\ Produced = Cycle\ Stock + Safety\ Stock - Current\ Inventory$$

Because the current inventory will on average be approximately equal to the safety stock, the quantity produced generally will be approximately the cycle stock.

If this model is used to order raw materials, generally there will be a lead time before the material is received. In this case, when the normal order interval begins enough material must be ordered to cover demand during the lead time as well as that needed to restore total inventory to the cycle stock plus safety stock target. Thus, the amount to be ordered is:

$$Order\ Quantity = DDLT + Cycle\ Stock + Safety\ Stock - Current\ Inventory$$

where DDLT = demand during lead time.

The current inventory will typically be approximately DDLT plus safety stock, so the amount ordered will be approximately the cycle stock.

Fixed Quantity Replenishment Model

Fixed quantity replenishment is an alternative to fixed interval, and is used when there is some benefit in buying or producing materials in specific quantities.

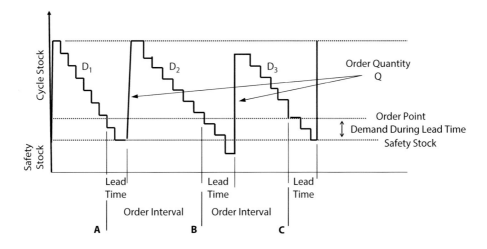

Figure E.4 **Inventory profile with fixed quantity replenishment.**

In the process industries, some materials are received in tank trucks or railcars, so transportation economics suggests buying in truck or railcar quantities, as is the case with adipic acid and diamine in our fiber process.

In the production process, there is often a specific campaign size that best balances changeover cost with inventory carrying cost, as determined by an EPQ (Economic Production Quality) calculation. That quantity could be used to replenish finished product inventory on a fixed quantity basis.

An inventory profile for a single material replenished using a fixed quantity model is illustrated in Figure E.4. Because the order quantity Q is already known, the question to be answered is when to place the next order. Whenever current inventory falls to or below the order point, a new order is placed. The time between orders can be variable: Order interval B–C is slightly shorter than interval A–B because demand D2 consumed some of the safety stock, so the replenishment quantity Q did not raise the inventory back to the target and we therefore hit the reorder point sooner on the next cycle. If, on the other hand, not all of the cycle stock had been consumed, Q would have raised the inventory above the cycle stock plus safety stock target. Thus, in this replenishment strategy the inventory peaks will float up and down, but the order point logic will ensure that the next order is always placed at the correct time.

Thus, in contrast with the fixed interval model, the interval here will vary, while the quantity ordered remains fixed. In the simplest case, the order point is calculated as:

$$Order\ Point = DDLT + Safety\ Stock$$

where DDLT = demand during lead time.

In a perfect world, with no safety stock, the new order would arrive just before a stockout would occur. However, in the real world, the new order may arrive late or the demand during the lead time might be greater than average, so we need safety stock to cover those situations. Thus, the order point is set so that on average, the new order will arrive just as inventory falls to the safety stock level.

This model is often called Continuous Review because it assumes that the inventory level is being continuously monitored, and the new order is placed immediately when the inventory falls to the reorder point. In many cases that is true, but in some situations inventory is not being checked continuously. Inventory status may be checked once per day, per week, or at some other frequency. The time between inventory examinations is called a review period. If the replenishment process includes a review period, it must be accommodated in the order point. In these cases the order point calculation is:

$$Order\ Point = DDRP + DDLT + Safety\ Stock$$

where DDRP = demand during review period.

In cases where the lead time is very long, this equation will result in a very large order point, which may alarm you. However, in either of these situations, current inventory includes not only the inventory on hand, but also inventory currently in transit and orders placed but not yet received.

Therefore, an order actually gets placed when:

$$\left(Inv\ on\ Hand + Inv\ in\ Transit + Unfilled\ Orders\right) < \left(DDRP + DDLT + Safety\ Stock\right)$$

And:

$$Peak\ Inventory = Cycle\ Stock + Safety\ Stock$$

$$Average\ Inventory = \frac{1}{2}\left(Cycle\ Stock\right) + Safety\ Stock$$

$$Cycle\ Stock = Q$$

Safety Stock

Safety stock is inventory carried to prevent or reduce the frequency of stockouts, and thus provide better service to customers. Safety stock can be used to accommodate:

■ Variability in customer demand or demand from downstream process steps (where demand history is used to set cycle stock).
■ Forecast errors (where forecasts are used to set cycle stock targets).
■ Variability in supply lead time or production lead time.

If all variations in demand and lead time are random and are reasonably normally distributed, the following calculations will result in appropriate safety stock levels. If not, they may still give some guidance, and are generally preferable to the rules of thumb sometimes recommended, that safety stock be set at 10%, or 20%, or 50% of cycle stock, or at some arbitrary number of days of supply.

Variability in Demand

To understand how we can avoid stockouts in the face of customer demand, which can vary up and down, a short lesson in statistics is in order. Figure E.5 is a histogram, a plot showing the number of cycles at which each demand range occurs. If we consider product A on baler number 2, with an average demand of 14 bales per weekly production cycle, the histogram shows how many weeks the true demand was within each range. The histogram shows that, for the 52 production cycles within a year, the demand was very close to the average for 18 of those weeks. In this plot, the width of each bar represents 1 bale; so on these 18 weeks (9 weeks plus 9 weeks), the demand was between 13 and 15 bales. It was somewhat higher, 16 bales, during 7 of the weeks, and 17 bales for 4 of the weeks. As the range of demand values goes higher, the number of weeks within that range decreases. There is a similar pattern on the other side of the average; for 7 of the weeks, the demand was 12 bales, and 11 for 4 weeks. This bell shaped curve is typical of many demand patterns.

Some products will have little variability and thus a very narrow histogram, while others will have higher variability and a wider histogram. The width of the curve, and the underlying variability, can be characterized by a statistical property called standard deviation and symbolized by sigma, σ. While the calculation of standard deviation is beyond the scope of this discussion, understanding σ

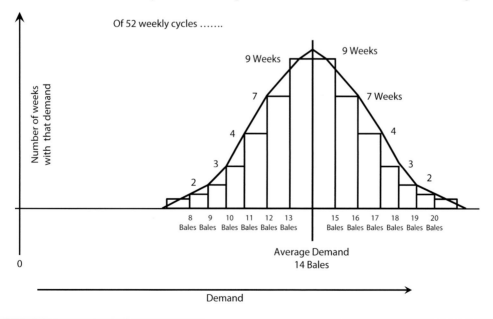

Figure E.5 A histogram of weekly demand.

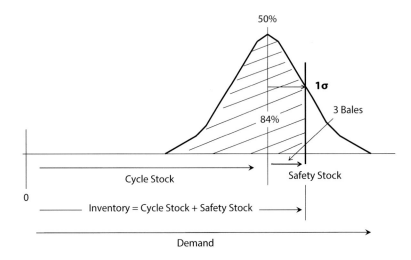

Figure E.6 **Safety stock equal to one standard deviation covers 84% of the cycles.**

can help us calculate how much safety stock we need to give us various levels of protection against demand variability.

If we carry no safety stock and have only the 14 bales of cycle stock, that will be enough to satisfy all demand for product A on half the cycles; half the time demand will be at 14 bales or less, and half the time greater. Therefore, with no safety stock, we will be vulnerable to stockouts on half the cycles. Statistics teaches us that if we carry extra stock equal to 1σ, that will be enough to cover demand on 84% of all cycles, as shown in Figure E.6. Sigma is 3 bales for product A, so if we carry 3 bales of safety stock in addition to the 14 bales of cycle stock, that should be sufficient to prevent stockouts on 84% of the cycles, about 44 weeks. If we carry safety stock equal to 2σ, that should cover 98% of the cycles, as shown in Figure E.7

Thus, the key to determining safety stock is deciding on the tolerance for stockouts, and then using that to determine how many sigmas of variability you need to cover. For example, if you decide that you can tolerate stockouts on no

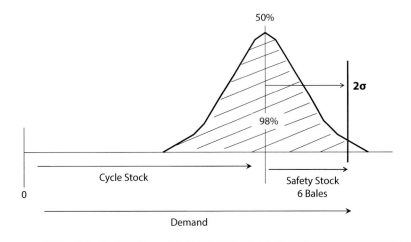

Figure E.7 **Safety stock equal to two standard deviations covers 98% of the cycles.**

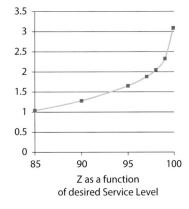

DESIRED CYCLE SERVICE LEVEL	Z FACTOR
84	1
85	1.04
90	1.28
95	1.65
97	1.88
98	2.05
99	2.33
99.9	3.09

Figure E.8 Relationship between service factor and service level.

more than 2% of the cycles, that sets the cycle service level goal at 98%, and we saw in Figure E.7 that that requires 2σ of safety stock, or 6 bales. The percentage of cycles you hope not to have stockouts is called cycle service level, and the number of sigmas required to achieve that is called the service level factor or the Z factor.

Thus, the general equation for safety stock required to cover demand variability is:

$$Safety\ Stock = Z \times \sigma_D$$

Figure E.8 shows the relationship between Z and service level. As can be seen, the relationship is highly non-linear: higher service level values, i.e., lower potential for stockout, require disproportionally higher safety stock levels. Statistically, 100% service level is impossible.

Typical service level goals are in the 90 to 98% range, but good inventory management practice suggests that rather than using a fixed Z value for all products, Z be set independently for groups of products based on strategic importance, profit margin, dollar volume, or some other criteria. Doing this will place more safety stock in those SKUs with greater value to the business, and less safety stock in the products believed to be less important to business success.

The above equation assumes that the standard deviation of demand is calculated from a data set where the demand periods are equal to the lead time or production cycle length. If not, an adjustment must be made to the standard deviation value to statistically estimate what the standard deviation would be if calculated based on the periods equal to the total lead time. As an example, if the standard deviation of demand is calculated from weekly demand data, and the lead time is 2 weeks, the standard deviation of demand calculated from a data set covering 2-week periods would be the weekly standard deviation times the square root of the ratio of the time units, or $\sqrt{2}$. Bowersox and Closs, in *Logistical Management*, use the term Performance Cycle (PC) to denote the total lead time. If we let T1 represent the time increments from which the standard deviation was calculated (1 week in the above example), PC to represent the total lead time or production cycle length, then

$$Safety\ Stock = Z \times \sqrt{\frac{PC}{T1}} \times \sigma_D$$

When procuring raw materials, the performance cycle includes the time to:

■ Decide what to order (order interval or review period).
■ Communicate the order to the supplier.
■ Manufacture or process the material.
■ Deliver the material.
■ Perform a store-in.

Inside our own manufacturing facility, the performance cycle includes the time to:

■ Decide what to produce.
■ Manufacture the material.
■ Release the material to the downstream inventory.
■ Return to the next cycle.
■ If we are carrying inventory in a finished product warehouse, and customers allow a delivery lead time greater than the time needed to deliver to the customer, then the remaining customer lead time can be subtracted from the Performance Cycle.

And with product wheels the performance cycle PC includes:

■ The product wheel time.
■ The time to get to the downstream inventory replenished by the wheel. There may be additional process steps between the wheel step and the next inventory storage location.
■ Any review period included in the wheel scheduling process.
■ Again, any excess customer lead time can be subtracted from the Performance Cycle.

The Performance Cycle can be considered the time at risk, i.e., the time between making a determination on how much to produce, and the time to make the next determination and have it realized.

If cycle stock has been calculated from historical demand, then the variance used in the safety stock calculation should be based on past demand variation. If forecasts are used to set cycle stock, then the thing requiring protection is forecast error. Standard deviation of forecast error would replace standard deviation of past demand in the safety stock formula, which would become:

$$Safety\ Stock = Z \times \sqrt{\frac{PC}{T1}} \times \sigma_{Fcst\ Err}$$

If there is bias in the forecast, it must be removed for the safety stock calculation to be valid. (Dealing with forecast bias is beyond the scope of this handbook.)

Variability in Lead Time

The above equation calculates the safety stock needed to mitigate variability in demand. If variability in lead time is of concern, that is, sometimes it takes longer to get back to that product in the cycle due to machine failures, temporary line upsets, or other problems, the safety stock needed to cover that is:

$$Safety\ Stock = Z \times \sigma_{LT} \times D_{avg}$$

where LT is lead time.

The average demand term (D_{avg}) is in the equation to convert standard deviation of lead time expressed in time units into production volume units (gallons, pounds, bales, etc.).

Combined Variability

If both demand variability and lead time variability are present, the safety stock required to protect against each can, under some conditions, be combined statistically to give a lower total safety stock than the sum of the two individual calculations. If demand variability and lead time variability are independent, that is, the factors causing demand variability are not the same factors influencing lead time variability, and if both variabilities are reasonably normally distributed, the combined safety stock is Z times the square root of the sum of the squares of the individual variabilities:

$$Safety\ Stock = Z \times \sqrt{\frac{PC}{T1}\sigma_D^2 + \sigma_{LT}^2 D_{avg}^2}$$

The idea behind this is that there is a low probability that you will see the highest demand and the longest production delay on the same cycle, that is, that you will be at the extremes of both variability curves at the same time.

If σ_D and σ_{LT} are not statistically independent of each other, that is, they are both influenced by the same factors, then this equation can't be used, and the combined safety stock is the sum of the two individual calculations:

$$Safety\ Stock = \left(Z \times \sqrt{\frac{PC}{T1}} \times \sigma_D \right) + \left(Z \times \sigma_{LT} \times D_{avg} \right)$$

Example—Cut–Bale 2 Safety Stock

As an example of the use of the above calculations, consider the inventory levels required in the storage being replenished by baler 2. Looking specifically at the inventory requirements for product A:

- Weekly demand = 14 bales
- Standard deviation of weekly demand = 3 bales
- Standard deviation of lead time = 0

Because the baler is running a 7-day cycle, and product A is being produced every cycle, the cycle stock for product A is 7 days' worth, 14 bales.

The lead time is stable and very predictable; baler reliability is poor, but high enough that the lead time never exceeds the seven days. So, with a standard deviation of lead time of zero, the safety stock requirement can be calculated from:

$$Safety\ Stock = Z \times \sqrt{\frac{PC}{T1}} \times \sigma_D$$

If 95% is the desired cycle service level (the business can tolerate stockouts of this product on no more than 5% of the replenishment cycles, slightly more than 2 per year), the Z value can be found in Figure E.8 to be 1.65. PC, the performance cycle, is the 7-day EPEI. T1, the time increments from which σ_D was calculated, is 7 days, so no correction to the standard deviation is necessary.

Thus,

$$Safety\ Stock = 1.65 \times \sqrt{\frac{7}{7}} \times 3\ bales = 5\ bales$$

(If the service level goal had been 98%, Z would be 2.05 and safety stock would be 6 bales.)

$$Peak\ Inventory = Cycle\ Stock\left(1 - \frac{D}{PR}\right) + Safety\ Stock$$

where D = 14 bales/week and PR = 125 bales/week

$$Peak\ Inventory = 14\ bales\left(1 - \frac{14}{125}\right) + 5\ bales = 18\ bales$$

$$Average\ Inventory = \frac{1}{2}\left(Cycle\ Stock\right)\left(1 - \frac{D}{PR}\right) + Safety\ Stock$$

$$Average\ Inventory = \frac{1}{2}(14\ bales)\left(1 - \frac{14}{125}\right) + 5\ bales = 12\ bales$$

When product A is being baled, it should make enough to bring the A inventory back up to the 18-bale level by the end of its production. This will require production of 14 bales plus any safety stock that was consumed, since approximately 2 bales will be consumed while it is being produced.

If there were variability in lead time, if baler equipment failures caused the 7-day cycle to vary, more safety stock may be required to meet the inventory performance goals. If, for example, lead time varied with a standard deviation of 1/2 day, or 0.07 weeks, the safety stock calculation would be as follows:

$$Safety\ Stock = Z \times \sqrt{\frac{PC}{T1}\sigma_D^2 + \sigma_{LT}^2 D_{avg}^2}$$

$$Safety\ Stock = 1.65 \times \sqrt{\left(\frac{7}{7}\right)3^2 + (0.07)^2\,14^2}$$

$$Safety\ Stock = 1.65 \times \sqrt{9 + 1} = 5.2\ bales \sim 5\ bales$$

Two things can be seen from this result. The first is that in this example the demand variability has the dominant influence on safety stock requirements: Its effect on safety stock is almost 10 times that of lead time variability. It is often the case that one factor or the other will dominate the calculation; it is important to recognize that, so that improvement efforts can be focused on the most appropriate things. In this case, if we decide to reduce the need for safety stock, it is far more productive to work on demand variability than on lead time variability. The second observation is that the influence of lead time variability is so small that safety stock requirements increase by a small fraction of a bale, so the effective increase on safety stock required is lost in the round off.

To summarize, given the parameters above, if the inventory for A includes 5 bales of safety stock, stockouts of product A should be expected to occur no more than 5% of the cycles, or during 2 to 3 of the weekly cycles each year. With the same service goals, stockouts of each of the other products should also be expected on 5% of the cycles.

If that number of stockouts is unacceptable, then higher service level goals should be set, resulting in higher Z values and more safety stock required. One key advantage of these calculations over arbitrary rules of thumb is that the trade-offs can be quantified, so the business can make informed decisions on how it wants to balance inventory cost against risk of stockouts.

Summary

The inventory for any specific product consists of cycle stock and safety stock. Cycle stock covers the cyclic nature of replenishment, and is determined by average expected demand, based on either demand history or a forecast. Safety stock is there to protect against stockouts in the face of variable supply, production, or demand.

Calculating inventory requirements using sound mathematical principles is always better than sizing inventory by tradition, gut feel, fixed percentages, or arbitrary days-of-supply targets. When arbitrary methods have been used, we frequently see that some inventories are much higher than needed while others are undersized. Use of a mathematical approach should both reduce the total inventory and improve customer delivery performance because we now have the appropriate inventory for each SKU in the product portfolio.

Appendix F: Product Wheels

Introduction to Product Wheels

Several of the improvement opportunities found from analyzing our current state VSM include the implementation of product wheels, a scheduling methodology that balances the production of all the products or materials made on a piece of equipment. It provides for production leveling, and is a way that many process operations achieve heijunka (the Lean term for production leveling). It also sets campaign cycles to the optimum balance between high inventory (the problem with long cycles) and high changeover costs (the problem with short cycles), and generally is very effective at reducing inventories.

It is generally recognized that manufacturing resources are utilized most effectively when production is leveled, when production is done at a uniform, constant, unvarying rate. This tends to maximize equipment utilization and labor utilization, and smooth out the requirement for raw materials and support facilities. For that reason, Lean principles teach us to level our production, a concept called heijunka.

Lean manufacturing also suggests that production be synchronized with the rate or rhythm of customer demand, that is, Takt. The concept is that if you can synchronize the steps in the manufacturing process to the rhythm of customer demand, you have a foundation in place to work toward smaller lot sizes, more continuous flow, and lower inventories, while satisfying customer demand with as little waste as possible. Unfortunately, for most operations, demand is not uniform at all. Therefore, if we have the resources, in labor, equipment, and raw materials, to produce to the demand peaks, much of that is idle in the valleys and is thus waste.

Heijunka challenges operations managers to balance the need to produce to varying customer demand with the need for production leveling. A number of companies in the process industries, including DuPont, Dow Chemical, Exxon Mobil, Apvion, and BG Products, Inc., have achieved this balance using the product wheel approach. These companies have found that product wheels not only level production, but also help to optimize product sequences and campaign lengths, and set inventory levels at the minimum required for smooth flow.

The benefits they've seen include both lower inventories and higher customer delivery performance because the inventory, although lower, is now in the right mix. In addition, usable production capacity generally increases because the regularity and predictability of the schedule reduces or eliminates unplanned schedule changes and the time they waste.

Product Wheels Defined

A product wheel (Figure F.1) is a visual metaphor for a structured, regularly repeating sequence of the production of all of the materials to be made on a specific piece of equipment, within a reaction vessel, within a process system, or on an entire production line. The overall cycle time for the wheel is fixed. The time allocated to each product (a "spoke" on the wheel) is relatively fixed, based on that product's average demand over the wheel cycle. The sequence of products is fixed, having been determined from an analysis of the path through all products which will result in the lowest total changeover time or the lowest overall changeover cost.

The spokes can have different lengths, reflecting the different average demands of the various products: high demand products will have longer spokes than lower demand products.

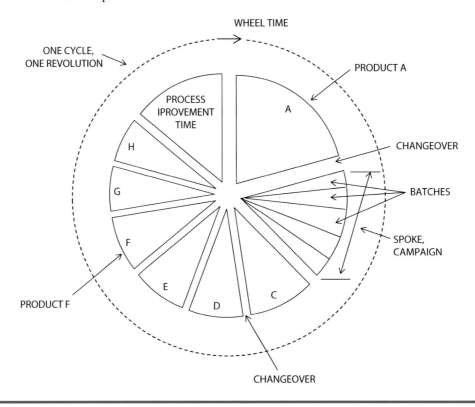

Figure F.1   The product wheel concept and terminology.

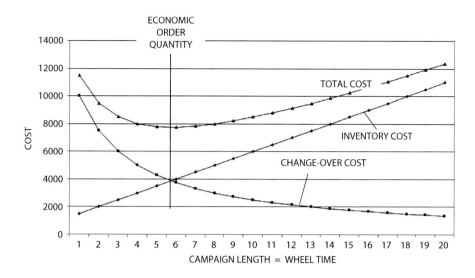

Figure F.2 Economic order quantity factors.

Product wheels support a pull replenishment model. That is, the wheel will be designed based on average historical demand or on forecast demand for each product, but what is actually produced on any spoke is just enough to replenish what has been consumed from the downstream inventory, in accordance with Lean pull principles. Thus the width of each spoke can vary from cycle to cycle based on actual demand, but the total wheel cycle time will remain fixed.

A number of factors can influence the determination of overall wheel time. If having the shortest possible lead time through the operation is paramount, then the shortest wheel time that allows for all required production and all necessary changeovers may be selected. If the lowest manufacturing cost is a key driver, then the wheel time can be calculated to give the best balance between changeover costs, which decrease with longer wheels, and inventory cost, which increases with longer wheels.

The graph in Figure F.2 illustrates this: Inventory carrying cost can be seen to rise with longer wheel cycles because there must be enough inventory of each product to carry you from one cycle to the next. Changeover costs exhibit the opposite effect. Longer wheel times result in fewer changeovers, while shorter wheels mean more changeovers within any time period. If maintaining the lowest practical operating cost is paramount to the business, then the wheel time can be set based on the optimum balance between these two costs, sometimes called Economic Order Quantity (EOQ) or Economic Production Quantity (EPQ).

The icon we use on future state VSMs to signify a process step scheduled by product wheels is shown in Figure F.3.

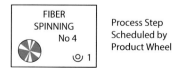

Figure F.3 VSM icon for a step with product wheel scheduling.

PRODUCT WHEEL ATTRIBUTES

- The overall cycle time is fixed

- The cycle time is optimized based on business priorities

- The sequence is fixed — products are always made in the same order

- The sequence is optimized for minimum changeover loss

- Spokes will have different lengths, based on the Takt for each product

- The amount actually produced can vary from cycle to cycle, based on actual consumption

- Some low-volume products may not be made every cycle

- When they are made, it's always at the same point in the sequence

- Make-to-order and Make-to-stock products can be intermixed on a wheel

Benefits of Product Wheels

Product wheels tend to level production as a natural behavior. And because production of a specific product on a specific cycle is based on inventory consumed during the last cycle, production is synchronized to the current Takt for that product.

The product wheel design methodology tends to drive scheduling to smallest practical campaign size, thus reducing inventory. The wheel design process finds the optimum overall cycle (EPEI) for the range of products made. It provides a clear understanding of trade-offs between changeover costs and inventory costs, so the cycle time can be optimized based on business priorities—shortest lead time or lowest cost operation. The technique also quantifies the benefits of further changeover improvement activities, so that the justification for capital improvements that would improve flexibility can be determined.

Product wheel design forces a structured analysis of various kinds of change-overs required, to optimize the sequence of production. People sometimes assume that they are following an optimized sequence, but a data-based analysis often reveals that improvements can be made.

A fixed, repeatable schedule gives a firm basis for determining inventory requirements. Because we know when we will be making a given product next, we can determine how much inventory will be needed to support demand until the next cycle. We also know how long we will be at risk to variation, and thus have a logical basis to calculate safety stock.

Product wheels offer a great deal of scheduling flexibility in that low volume products can be programmed to be made at a frequency less than every cycle, and Make-to-Order products can be made on the same wheel with Make-to-Stock products.

Product wheels add predictability to high mix operations, i.e., operations that make a large number of products. Everyone associated with an operation scheduled by product wheels knows what is going to be produced and at what time. Any special tools, materials, or personnel required for changeovers can be scheduled in advance. If the startup of the next material places additional loading on the test lab, this can be planned for.

This regularity, predictability, and rhythm—what Ian Glenday calls the "Economies of Repeatability"—is perhaps one of the greatest benefits that wheels bring to the operation. This regularity leads to a dramatic reduction in unplanned schedule changes, and therefore increased usable capacity. Moreover, because people are no longer spending most of their time firefighting, they now have the bandwidth to deal with any real schedule emergencies that arise.

To summarize, the reduction in all the normal chaos and firefighting leads to more usable capacity; the fixed, repeatable cycle allows inventory requirements to be more accurately known and generally reduced; and because the inventory is in the right mix, customer delivery performance improves.

PRODUCT WHEEL BENEFITS

1. Leveled production
2. Improved changeovers via optimized sequences
3. Increased usable capacity
4. Optimized campaign lengths
5. More realistic inventory target setting
6. Reduced inventory
7. Improved delivery performance
8. A higher degree of regularity and predictability in operations
9. A credible mathematical basis to support decision making

Product Wheel Applicability

Product wheels may be applied to an entire production line, such as a salad dressing bottling line, a synthetic fiber spinning operation, or a frozen pizza manufacturing/packaging line. Or they may be applied to a single large piece of process equipment, such as a plastic pellet extruder, a paper forming machine, or a resin reactor used in paint manufacture.

Process Improvement Time

There is often more time within a wheel revolution than is needed for production and changeovers. This time can be extremely valuable to the operation for:

■ Preventive maintenance tasks
■ Equipment modifications
■ New product development trials
■ New product qualification runs
■ Operator training
■ Kaizen events to implement 5S practices
■ Kaizen events to reduce changeover time and cost, using Single Minute Exchange of Dies (SMED) or other techniques

Referring to this available time as "Process Improvement Time" (PIT Time) tends to highlight how valuable it can be, and creates a mindset that it should not be wasted.

Summary

Product wheels are a time tested and proven methodology for resolving a number of operations management issues in an integrated, holistic way. DuPont has been using wheels beneficially for more than 20 years, and Dow Chemical more than 10. Optimizing the product sequence generally reduces changeover losses and thus increases usable capacity. The predictability and stability wheels bring to the operation reduce the frequency of illogical schedules changes, and thus also contribute to reduced losses and increased capacity. Right-sizing the inventory to support the wheel schedule generally reduces inventory and puts it in the right mix balance so that customer service improves. So it's not surprising that as more and more companies learn about product wheels, they adopt the concept as a way to balance the production of a wide variety of product types.

A thorough walk-through of the steps in wheel design, implementation, and operation can be found in *The Product Wheel Handbook—Creating Balanced Flow in High Mix Process Operations*, P.L. King and J.S. King, Productivity Press, 2013.

Appendix G: Additional Reading

Arnold, J. R. Tony, and Stephen N. Chapman, *Introduction to Materials Management*. Upper Saddle River, NJ: Pearson Prentice Hall, 2004.

Blackstone, John H. Jr. (Editor), *APICS Dictionary*, 12th ed. Chicago, IL: APICS, the Association for Operations Management, 2008.

Bowersox, Donald J. and David J. Closs, *Logistical Management*. New York: McGraw-Hill, 1996.

Chopra, Sunil, and Peter Meindl, *Supply Chain Management—Strategy, Planning, & Operation*. Upper Saddle River, NJ: Pearson Prentice Hall, 2007.

Floyd, Raymond C., *Liquid Lean*. New York: Productivity Press, 2010.

Ford, Henry, *Today and Tomorrow*. Portland, OR: Productivity Press, 1988.

Glenday, Ian and Daniel Jones, *Breaking Through to Flow: Banish Firefighting and Produce to Customer Demand*. Herefordshire, U.K.: Lean Enterprise Academy, Ltd., 2006.

Goldratt, Eliyahu M., *Theory of Constraints*. Great Barrington, MA: North River Press, 1990.

King, Peter L., *Lean for the Process Industries—Dealing with Complexity*. New York: Productivity Press, 2009.

King, Peter L., and Jennifer S. King, *The Product Wheel Handbook—Creating Balanced Flow in High-Mix Process Operations*. New York: Productivity Press, 2013.

Liker, Jeffrey K., *The Toyota Way*. New York: McGraw-Hill, 2004.

Liker, Jeffrey K., and David Meier, *The Toyota Way Fieldbook*. New York: McGraw-Hill, 2006.

Ohno, Taiichi, *Toyota Production System: Beyond Large Scale Production*. New York: Productivity Press, 1988.

Panchak, Patricia, "Leveling and pull streamline production in process industries." *Target Magazine*, First Issue, 2009.

Rother, Mike, and John Shook, *Learning To See*. Cambridge, MA: The Lean Enterprise Institute, 2003.

Rummler, Geary A., and Alan P. Brache, *Improving Performance—How to Manage the White Space on the Organization Chart*. San Francisco, CA: Jossey-Bass, 1995.

Schonberger, Richard J., *World Class Manufacturing, The Lessons of Simplicity Applied*. New York: The Fee Press, 1986.

Shingo, Shigeo, and Andrew P. Dillon, *A Revolution in Manufacturing: The SMED System*. Cambridge, MA: Productivity Press, 1985.

Smalley, Art, *Creating Level Pull*. Brookline, MA: The Lean Enterprise Institute, 2004.

Smith, Wayne K., *Time Out*. New York: John Wiley & Sons, 1998.

Umble, Michael, and Mokshagundam L. Srikanth, *Synchronous Manufacturing, Principles for World Class Excellence*. Cincinnati, OH: South-Western Publishing Co., 1990.

Womack, James P., Daniel T. Jones, and Daniel Roos, *The Machine that Changed the World*. New York: Macmillan Publishing Company, 1990.

Index

About the Authors

Jennifer S. King is an Operations Research Analyst with MCR LLC, analyzing operational impacts of emerging FAA technologies and developing cost and performance models to support airline investment decisions. Prior to that, she spent five years with the Department of Defense developing discrete event simulation models to assist the army in setting reliability requirements for new platforms, and analyzing performance of weapon systems alternatives. Her prior publishing experience includes editing textbooks and developing mathematics problems and solutions for ExploreLearning. She is the co-author of *The Product Wheel Handbook—Creating Balanced Flow in High Mix Process Operations* (Productivity Press, 2013).

Jennifer has degrees in Mathematics and Psychology from the University of Virginia, and a master's degree in Operations Research from the University of Delaware. She is a member of INFORMS.

Peter L. King is the president of Lean Dynamics, LLC, a manufacturing improvement consulting firm located in Newark, DE. Prior to founding Lean Dynamics, Pete spent 42 years with the DuPont Company, in a variety of control systems, manufacturing systems engineering, Continuous Flow Manufacturing, and Lean Manufacturing assignments. That included 18 years applying Lean Manufacturing techniques to a wide variety of products, including sheet goods like DuPont™ Tyvek®, Sontara®, and Mylar®; fibers such as nylon, Dacron®, Lycra®, and Kevlar®; automotive paints; performance lubricants; bulk chemicals; adhesives; electronic circuit board substrates; and biological materials used in human surgery. On behalf of DuPont, Pete has consulted with key customers in the processed food and carpet industries. Pete retired from DuPont in 2007, leaving a position as Principal Consultant in the Lean Center of Competency. Recent clients have included producers of sheet goods, lubricants and fuel additives, vitamins and nutritional supplements, and polyethylene and polypropylene pellets.

Pete received a bachelor's degree in Electrical Engineering from Virginia Tech, graduating with honors. He is Six Sigma Green Belt certified (DuPont, 2001), Lean Manufacturing certified (University of Michigan, 2002), and is a Certified Supply Chain Professional (APICS, 2010). He is a member of the Association for Manufacturing Excellence, APICS, and the Institute of Industrial Engineers. He served as president of IIE's Process Industry Division in 2009–2010.

Pete is the author of *Lean for the Process Industries—Dealing with Complexity* (Productivity Press, 2009), and several published articles on the application of Lean concepts to process operations. He is the co-author of *The Product Wheel Handbook—Creating Balanced Flow in High Mix Process Operations* (Productivity Press, 2013). He has been an invited speaker at several professional conferences and meetings. He has presented seminars and taught courses across the globe on the application of Lean concepts to process operations.